Sa...

Travel
Companion

Safe Air Travel Companion

DAN McKINNON

McGraw-Hill

New York   Chicago   San Francisco   Lisbon   London
Madrid   Mexico City   Milan   New Delhi   San Juan   Seoul
Singapore   Sydney   Toronto

Cataloging-in-Publication Data is on file with the Library of Congress.

McGraw-Hill

A Division of The **McGraw·Hill** Companies

1 2 3 4 5 6 7 8 9 0   DOC/DOC   0 7 6 5 4 3 2 1

ISBN 0-07-139918-6

The sponsoring editor for this book was Shelley Carr, the editing supervisor was Caroline Levine, and the production supervisor was Pamela Pelton. It was set in Veljovic Book by McGraw-Hill's Professional Book Group composition unit at Hightstown, NJ.

Printed and bound by R.R. Donnelley & Sons Company.

Previously published by House of Hits Publishing.

McGraw-Hill books are available at special quantity discounts to use as premiums and sales promotions, or for use in corporate training programs. For more information, please write to the Director of Special Sales, McGraw-Hill Professional, Two Penn Plaza, New York, NY, 10121-2298. Or contact your local bookstore.

This book is printed on recycled, acid-free paper containing a minimum of 50% recycled, de-inked paper

---

Information contained in this work has been obtained by The McGraw-Hill Companies, Inc. ("McGraw-Hill") from sources believed to be reliable. However, neither McGraw-Hill nor its authors guarantee the accuracy or completeness of any information published herein and neither McGraw-Hill nor its authors shall be responsible for any errors, omissions, or damages arising out of use of this information. This work is published with the understanding that McGraw-Hill and its authors are supplying information but are not attempting to render engineering or other professional services. If such services are required, the assistance of an appropriate professional should be sought.

CONTENTS

Preface

"WE'RE AT WAR." SO DECLARED PRESIDENT George W. Bush upon first learning that an American Airlines B767 and a United Airlines B767 had smashed into the World Trade Center towers the morning of September 11, 2001.

More catastrophe was yet to come: Another American B757 knifed into the Pentagon and a second United B757 nosedived into a Pennsylvania farm field. The total dead among the airline passengers and crew was 266, including the 19 hijackers.

In addition, more than 3,000 innocent people died in the fire and collapse of the World

Trade Center towers, and an additional 189 victims perished in the Pentagon crash.

This suicidal terrorist attack, which turned airliners into cruise missiles, has changed totally the way the world operates.

After the attacks, airliners in the United States were instructed to land immediately at the nearest suitable airport; they stayed there for at least the next 48 hours—or longer depending upon security measures implemented at their next intended destination.

The skies over America were silent.

Up until September 11 there were an average of 31,941 airline flights scheduled daily by the major carriers in the United States with an average of 1,914,600 daily travelers. That worked out to about 700 million people who were flying annually. So the chances of being hijacked were minuscule by comparison to the numbers of people traveling.

But once airlines started flying again, in the days after September 11th, fear so gripped the nation that for a while there were more crewmembers than passengers aboard most domestic flights.

Prior to September 11th a hijacking generally meant an extortion effort or a quick trip to Cuba.

Americans were now confronted with the realization that war and terrorism had hit our homeland for the first time. We had watched it in Israel and in Europe but always felt we were immune in the United States, the greatest and strongest country the world has ever known.

With the news media so numerous and diversified and hungry for any dramatic news to report, a terrorist attack of any kind becomes instant worldwide headlines. September 11 scared off travelers, plunging the helpless airlines into economic chaos, with cataclysmic ripple effects throughout the economy of our country and shudders throughout the rest of the world.

Despite these attacks, the odds of being a victim of a terrorist attack or hijacking are about one in eleven million. Nonetheless, with terrorism on the rise it is smart to prepare yourself mentally to avoid such dangers and think about how you would deal with them.

Americans have a built-in travel lust. They will demand better security but will not give up their right to see what is going on in the world firsthand.

Terrorism is just one more risk of traveling you can't completely control.

What terrorists want you to do is cancel your trip. Their aim is to force you to

- ○ Avoid areas they make a threat or danger.

- ○ Seek the security of only your own home.

- ○ Isolate yourself from the rest of the world.

- ○ Crawl back into your shell.

And if you let them —they've won. They've accomplished what they set out to do. They're controlling the world. They, not you, are determining what you are doing.

Your freedom has been denied.

And your lifestyle will be greatly diminished.

President George W. Bush has promised "whether we bring our enemies to justice or bring justice to our enemies, justice will be done."

In the meanwhile there are two choices: cancel your trip and stay home forever, or learn how to deal with the potential problems.

Safe Air Travel Companion is about how to deal with those potential terrorist problems so you can continue to travel.

Dan McKinnon

Introduction

THERE ARE NO ABSOLUTE GUARANTEES AGAINST being involved in a hijacking or becoming a hijack victim.

There are no absolute guarantees of not being involved in a car accident, falling off a ladder or down a flight of stairs, slipping in a bathtub, or being struck by lightning.

This book is designed to give a brief overview into

- The problems of hijacking and terrorism

- How remote the chances are of being hijacked

○ Steps to take to minimize the possibility of it happening to you

○ What to do if it should happen to you—some information to help you survive the ordeal

This book is full of suggestions and checklists, but it doesn't have all the answers to terrorist hijackings. We are playing a cat-and-mouse game with hard-core committed terrorists who are ready to commit suicide. They are always thinking of new and clever countermeasures for every defensive measure taken to ensure passenger safety.

Terrorism and hijacking are form of warfare. But we can't allow them to paralyze our lives. They have become one more eventuality of life we need to prepare for, but we can't stop living and we don't want to stop traveling. *Safe Air Travel Companion* is offered as a guide to prepare ourselves for things as best we can and get on with life.

About the Author

DAN MCKINNON IS AN EXPERT IN COMMERCIAL aviation travel and survival.

He currently is the owner and president of North American Airlines based at JFK International Airport. North American is a worldwide large jet charter and scheduled airline flying B757 and new generation B737-800 aircraft.

From 1981 to 1985 he served as Chairman of the Civil Aeronautics Board. He was appointed by President Ronald Reagan and oversaw implementation of airline deregulation during the tumultuous period of bankruptcies and adjustment from a government-regulated industry to one controlled by the marketplace.

McKinnon played a key role in U.S. international aviation policy and negotiations for air-route agreements with countries around the world.

On December 31, 1984, McKinnon oversaw the shutdown of the 46-year-old Civil Aeronautics Board in accordance with the President's wishes—the first government regulatory agency ever closed.

McKinnon formerly owned and operated two radio stations in San Diego for 23 years and spent 4 years as publisher of a newspaper in La Jolla, California.

A former Navy pilot, who specialized in rescue efforts and techniques, McKinnon holds the Navy peacetime helicopter rescue record with 62 air-sea saves. Those exploits are recounted in *Rescue Pilot*.

In the middle 1980s he did special projects for the Central Intelligence Agency.

As part of his military training, and continuously since that time, he has done extensive study in survival, captive, and POW situations.

Safe Air Travel Companion

Chapter 1

Why Terrorism?

TERRORISM TODAY DERIVES FROM SEVERAL POLIT-
ical or religious philosophies.

It used to be that some of the smaller countries or individuals who exported terrorism
simply could not compete in the marketplace of power or with sophisticated
weapons to impose their causes or ideas on
others. Besides, weapons of mass destruction as well as increasingly lethal conven-
tional armaments made regular warfare too
costly.

But with ideological kamikazes willing to
sacrifice their lives to destroy so many others no target is safe, causing terrorism to
take on a new dimension.

Through violence terrorists hope to promote fear and demoralize America to gain leverage over our democratic government and to destroy civilized governments around the world. Their aim is to goad democracies to abandon the rule of law and overreact to fight terrorists on their own level. Indeed, they see a response so repressive that citizens in the free world will be alienated from their governments and cause unrest, dissidence, and disloyalty.

The terrorist appetite for publicity and recognition is a driving force. They are event-oriented. That is why airliners, airports, and significant landmarks are prime targets. They want our American free society to feel vulnerable.

People have a natural fear of flying, and that, plus the fact that flying is such a popular mode of transportation, allows everyone to relate to what goes on in an airplane.

Any dramatic embellishment of fear—like a terrorist attack—magnifies the impact of terrorists out of proportion to the threat. Radical terrorists are "sleepers" waiting to be activated to commit some dastardly deed, even if their own life is part of the cost.

According to the Center for Intelligence Studies, tens of thousands of terrorist operatives and auxiliaries are scattered in dozens of countries throughout the world. That such a huge terrorism presence exists may reflect a need to re-examine immigration polices, not just in the United States but throughout the Western world.

Twentieth Century International Hijacking

Serious international hijacking that captured world attention began July 23, 1968, with the hijacking of an El Al flight departing Paris for Tel Aviv with a stopover in Rome. The Boeing 707 was commandeered by three young Palestinians. The plane was forced to land in Algeria.

This was no simple extortion—no homesick passengers and no freedom-seeking refugees. It was a cold, calculating political effort to create problems for Israel and gain publicity for the PLO and its cause.

El Al immediately changed the way it did business.The Israeli airline structurally modified its fleet of aircraft to withstand bomb blasts in certain areas, added air locks as an in-flight way of disposing of bombs found aboard, beefed up the cockpit areas with bulletproof barriers as well as a double-door system, and added armed guards.

At Lod Airport outside of Tel Aviv, El Al officials instituted preliminary screening before automobiles were allowed near the terminal building. Baggage left unattended for more than a moment was whisked to underground blast shelters away from the terminal building and had to be reclaimed by the careless passengers, who would probably miss their flight. Among other precautions, plainclothes security agents mixed with passengers.

A form of selective profiling was adopted with extensive interviews of each passenger boarding an El Al flight. Anyone suspected of trouble making is screened out and kept off the flight.

Although the added precautions were time consuming for passengers, today El Al is considered the safest airline in the world in terms of terrorist attack.

In most terrorist hijackings in the past, relatively few passengers were singled out for death. Those selected for such punishment by terrorists stood out for some reason as being representative of the ideology the terrorist were opposing.

However, today's callous terrorists are not bothered by murder.

They want to impose their will forcefully on others. The idea is to create an atmosphere of fear and feeling of helplessness among their victims and their fellow countrymen. They also design their acts of terror so anyone who witnesses or is aware of them will say to themselves: "That could have been me."

The terrorists' goal is to paralyze and strike fear into the heart of anyone who opposes their viewpoints. There is nothing rational in their behavior. Hostages are just pawns in the bigger game of life as far as these terrorists are concerned, and if random victims die, all the better for creating fear.

Former CIA Director William J. Casey, who had great foresight, put it succinctly and prophetically 15 years ago when he stated:

> Whatever his specific political program, the terrorist always pursues one general goal— to fix in the public consciousness a sense of the terrorist's omnipotence and the public's helplessness.

> The terrorist, in short, has declared war on the mind. It is the impression of being everywhere and nowhere, of striking with impunity at whomever and whatever he will, that gives the terrorist his real power. To do this, the terrorist takes advantage of the very civilization he seeks to destroy.

> The terrorist depends upon two factors for success in conducting his war on the mind. Both of these factors, ironically, are found only in the urban centers of open societies like ours and those of our friends around the world.

> The first, and most important of these, is coverage by the media. In this decade, more people can be addressed by newspaper, television, radio and magazines than ever before in history. What is more, the media is so effective that millions of people may learn of a terrorist attack that has taken place half a world away in a matter of minutes—or at most—hours.

It is no accident that the vast bulk of the most heinous terrorist murders, bombings and hijackings take place—not in isolated villages in Africa or among remote Asian tribesmen—but in cities that possess excellent communication links with the rest of the world. The terrorist hopes that his deeds will be bannered on the 6 o'clock news throughout the developed world, will be commented on at length in the world's leading newspapers, and perhaps become the subject of everyday conversation.

People will ask: "When will the next attack occur?" "Where will the terrorists strike next?" "Who will be the victim?" Such uncertainty has a numbing effect on millions of people who expect to stand by helplessly to witness the next outrage. Or, perhaps, they expect to be killed or maimed in the next savage terrorist attack.

Even if an attack fails, the terrorist will nonetheless gain the maximum psychological impact of his deed by a bold public threat. The IRA in a public notice about a failed assassination attempt told Mrs. Thatcher: "This time you were lucky. But you have to be lucky all the time. We only have to be lucky once." Clearly, the threat and the uncertainty is a powerful weapon.

In the case of the failed assassination attempt the terrorists nonetheless succeed because their objective was the creation of an atmosphere of fear and uncertainty.

The second factor that aids terrorists in their campaign is the nature of modern urban society. The concentration of population offers anonymity to the terrorist.

Weapons and money can be obtained through an infinite number of channels, thus preserving the terrorist's operation security. The variety and efficiency of transportation enhances the terrorist's mobility. Moreover, industrialized societies have more vulnerable high-value targets—such as computer centers, airlines, factories, shopping arcades and even apartment complexes. It is also no secret that democratic societies provide more opportunities for a terrorist, and certainly more room to maneuver.

Unlike Russia or other closed societies that require internal passports and have frequent police checks on visitors and travelers, Western societies have only the lightest checks on movement. The ability to live where one pleases and to associate with whomever one chooses aids the terrorist in his operations. Thus, Western democracies by their very nature are particularly vulnerable to terrorist attacks.

Terrorism really doesn't need to be explained —only answered.

Chapter 2

Putting Hijacking in Perspective

UNITED STATES SCHEDULED AIRLINES TRANSPORT almost two million passengers daily—more than the airlines of any other nation—and nearly as many as will be flown by the rest of the world's airlines combined. Plus there are hundreds of charter flights daily by smaller airlines flying large jet airliners, with the same safety requirements as the large airlines. (The charter flights are not as symbolic a target as a major airline, and some fly the latest state of the art equipment. As a result, some feel they are less of a terrorist target.)

The 31,941 daily flights of the nation's airlines take place from almost 400 airports in the United States and about 100 other points throughout the world. The service is being provided by a fleet of more than 7000 high-technology aircraft, which represent an investment of about $140 billion.

The high visibility and mobility of air transportation make airlines targets of would-be sabotage or hijacking, prompted by terrorism or other motives.

The occurrence rate of hijacking is very low when you are looking at the millions of flights operated annually. But any actual or attempted hijacking creates fear and threatens safety—and in air transportation, safety is of paramount importance. That is why the airline industry and the government have worked together to develop what they thought were extensive measures to counter hijacking and other crimes against air transportation.

The world's first hijacking of a commercial aircraft occurred in 1931 on a domestic flight in Peru. The first hijacking in the United States occurred 30 years later. But hijacking remained a comparative rarity until a flurry of hijackings began in 1968, with many planes forced from the United States to Cuba.

From 1945 to 1952, the vast majority of hijackers were Eastern Europeans who wanted to escape communist domination. There was little worldwide concern or focus of attention on this problem because the Western world shared the same political ideas as the hijackers. Most hijackings were for extortion or a desire to cross an international border.

In the first 41 years of U.S. airline operations, ending with 1967, there were only nine hijackings.

Then, in 1968 there were 16 hijackings. In 1969 there were 40. In 1970 and 1971 there were 25. In 1972 there were 27. That made a five-year total of 133 hijackings in the United States. During that same period in the rest of the world, there were 165 hijackings.

Passengers and government officials became alarmed.

One hundred percent screening of U.S. passengers and their carry-on luggage was the response. It began in January 1973 at almost every airport in the United States.

Screening had a dramatic impact on limiting hijackings, except in 1980, when there were 21 cases; and in 1983, when there were 16 incidents. These hijackings were primarily the result of the Mariel boatlift that brought 125,000 Cuban refugees, including released criminals, to the United States in the spring of 1980. Some of the homesick and disillusioned refugees wanted to return to Cuba, and a free hijacked airplane flight was the easiest way.

The U.S. hijack record since the initiation of passenger screening looks like this:

Year	Hijackings
1973	1
1974	3
1975	6
1976	2
1977	5
1978	8
1979	11
1980	21
1981	7
1982	9
1983	16
1984	5
1985	4
1986	3
1987	4
1988	1
1989	1
1990	2
1991	1
1992–2000	0
Total	114

During the same period (1973–2000), there were 446 hijackings in the rest of the world.

Domestic hijackings stopped after 1991, in part due to U.S. screening techniques that were adopted by airlines. The last hijacking of a U.S. carrier was on February 10, 1991. A Southwest Airlines flight from Oakland, California to Austin, Texas was hijacked. The FAA wouldn't say whether an air marshal was aboard. The hijacker, who was attempting to divert the plane to Cuba, was arrested when the pilot landed the aircraft in San Diego.

In 1982, Cuban authorities announced aircraft hijackers to that country were getting tough prison sentences of 12 to 20 years. Cuban hijackings then took a dramatic drop.

With threats of capture and punishment at both ends of the hijacking trip to Cuba, such hijackings virtually disappeared. This was important because of the "copycat" phenomenon associated with hijacking. When Cuban-bound hijackings slackened, hijackings prompted by other reasons also diminished.

Other than Cuban desires to return home, the other big reason for U.S. airline hijacking was extortion.

The most celebrated case was D.B. Cooper, who on November 24, 1971, demanded $200,000 and four parachutes. While airborne he strapped on a parachute, stuffed his clothes with money and left a Northwest Airlines 727 through the lowered rear stairs at night. He bailed out over the Northwest part of the United States. Cooper was never seen or heard from again despite an intensive ground search.

To guard against that type of escape again, the stairs at the rear of all 727s were immediately modified so they could not be opened in flight.

The death penalty for hijacking a plane was adopted in 1974. It is now one of the few federal crimes punishable by death.

The passenger screening process was tough and demanding when first implemented compared to no oversight at all. But, when it was instituted, the majority of U.S. citizens were happy to give up some aspects of their right to privacy for the common good. There were some court challenges to this intrusion, but they were eventually denied and screening of all passengers remains the law of the land.

However, screening has also been under constant criticism for not being thorough enough, and at various times it has been subject to debate about being federalized.

Another type of passenger screening instituted was an FAA-devised hijacker behavioral passenger profile.

The profile, used to indicate whether or not a person was a hijack risk, was implemented prior to 1973 and proved remarkably effective when applied with proper follow-through procedures. It became an effective tool to keep potential hijackers off airplanes.

While they were checking in and boarding the aircraft, most airlines screened passengers with a profile that consisted of 17 negative points and 37 positive facets. The carriers concentrated on individuals with a negative profile and then made judgments on allowing those passengers to board the aircraft, based largely on an interview with the passenger.

Human error, improperly applied profiles, or lack of follow-through were the only ways a hijacker could slip through this observation net.

With computer systems in airline ticketing and improvements in high technology, the means exist to take human error out of applying most of the profile data to detect potential hijackers. The profile is built into the computer systems for automatic comparisons, but was not widely used until the twenty-first century.

With the dramatic rise in airline passengers and the lack of hijacking attempts use of the profile screening had become stagnate.

The profile screening is important because more than half of all U.S. hijackers carried no gun or other weapons. They made scary, verbal threats that were too risky to challenge once a plane was airborne, since passenger and aircraft safety come first.

In almost all hijackings—79 percent—the hijackers boarded as ticket-carrying passengers. That's why profiles are so important.

There have been only two U.S. hijackings involving firearms. There have been 22 instances of hijackers threatening to use alleged flammable liquids in the hijacking. All were by ticketed passengers.

In 54 percent of all hijackings, no weapons were involved. The hijackers claimed that weapons, liquids, or explosives did exist and would be used, to threaten the safety of the air crews and passengers.

With this in mind, 72 percent of all hijackers met the FAA hijacker profile—so it's easy to see why the emphasis was placed on the profile as a means to detect and stop this form of terror in the skies.

There were intense efforts to upgrade the use of the profile as a further deterrent. Efforts included such things as computer-designed hijacker behavioral profiles, a flexible set of criteria against ticket costs, miles to be flown, most frequent hijacker departure and scheduled arrival times at airports, and then matching that information with tickets purchased and other factors.

However, a sensitivity to concern about discrimination prevents profiling by race, ethnic background, or religious affiliation.

Profiling policy could have a dramatic impact in halting international terrorism and hijacking as well if foreign countries would make a strong profile effort. In many foreign countries, nationals of that country are required to do basically all the work of servicing and boarding U.S. and other foreign airlines arriving or departing that country. There are restrictions on U.S. carriers as to how many U.S. citizens can work for a U.S. airline overseas. Such work conditions are spelled out in aviation bilateral treaties, and it is a distressing fact for many

U.S. international carriers. It has a negative impact on their ability to screen their own passengers.

In the United States, airlines were told by the government to assume security responsibilities, including screening that normally would be considered traditional government public safety responsibilities. That way the airlines and traveling public had to absorb the security cost—not the government.

However, the threat of being screened and the fear of related security measures in effect have eliminated thousands of potential crimes and hijackings. There are few who will disagree with the benefit of security screening or advocate doing away with it.

The first reported case of a bomb explosion aboard a commercial aircraft was on a domestic flight in the Philippines in 1949, and the first such incident on an airline in the United States came six years later. There have been 20 crimes of this type reported in U.S. airline history and about 60 involving foreign airlines.

The security programs of both the airlines and airport operators cover far more than screening. The airlines have long had special procedures designed for cargo and checked baggage.

With cargo, for example, high-quality identification is often required from persons tendering shipments to an airline. In cases where the shipper is not well known to the carrier, the tendered shipment can be subject to especially detailed inspection.

To achieve higher security in handling checked baggage on certain flights, the airlines have had special checked baggage screening procedures in effect for over 25 years.

The airlines also protect parked aircraft, control access to aircraft, and provide a safe environment during flights. The security responsibilities of airport operators range from the identification of airport employees to the security patrols of airport roadways and parking areas to the perimeter fencing.

Airline flight and cabin crews undergo training to help them cope with a hijacking incident. But the terrorist aspect now requires additional training.

Once a hijacking is in progress, both the FAA and FBI have clearly defined areas of responsibility.

Highly trained federal air marshals are used on selected U.S. airline flights in high-risk areas in the United States and abroad.

Intelligence

A free society's lack of imagination of what a terrorist mind could dream up to harm innocent people invites attack.

Lack of intelligence creates vulnerability as well. It is believed there is a network of thousands of terrorist operatives in dozens of countries plus thousands of sympathizers. Our CIA has about 18,000 employees of which only around 800 are actual human spies or case workers as the agency refers to them.

Terrorist attacks have jolted the U.S. intelligence community and law enforcement agencies into action. There's a renewed interest and dedication based on the realization that their work is important and can have an effect on everyday life here in America.

Coordination between FBI and CIA is standard. Additionally the Defense Intelligence Agency and National Reconnaissance Office coordinate their activities as part of the intelligence establishment's endeavor to unify efforts.

Local law enforcement and intelligence gathering mechanisms are also integrated into the federal intelligence gathering programs. Regional FBI offices and agents work closely with local law enforcement organizations.

It is important that local law enforcement be involved because they can spot irregular activities that form patterns and pieces of the overall mosaic that establishes a pattern of terrorist activity.

Details of system integration cannot be disclosed for obvious security reasons. However it is known that the CIA is making renewed efforts to get case workers to develop agents within terrorist organizations and in sympathizer terrorist countries.

Such penetration is not an overnight achievement. It takes years for a CIA operative in a strange land to gain the confidence of someone and then discover the key that motivates them to share sensitive secrets about their terrorist operations to their enemy. But without spies sniffing around there is no finding out what is going on in the minds of unsavory people.

Restrictions placed on the CIA to hire model citizens abroad as agents are being relaxed. The agency and administration are accepting the fact that agents who do not fit the mold can provide helpful intelligence.

All government agencies make a concentrated effort to piece together bits of intelligence quickly enough to intercept and thwart any terrorist intentions.

Every citizen is aware of individual responsibility as well. Any information on terrorist threats should be reported to the FBI or State Department at the following phone numbers, addresses, or Web sites:

Telephone 1-800-437-6371 (U.S. only) or 1-866-483-5137

Mail Rewards for Justice
P.O. Box 96781
Washington, DC 20090-6781

Internet www.dssrewards.net or
www.ifccfbi.gov

E-mail mail@dssrewards.net

Chapter 3

What the Government Is Doing to Protect You

Our government is reacting quickly to terrorism in America.

In the process required to maintain our basic way of life, citizens will have to give up some of their rights and freedoms. There is no choice.

Some of the immediate actions being taken by the government for the traveling public's safety include increasing the number of air marshals, new standards for passenger screeners, positive baggage match, increased cockpit security, new restrictions on what passengers can carry onto the aircraft, and changes in the type of cargo and its screening.

○ The number of air marshals has been dramatically increased and supplemented with personnel from other federal law enforcement agencies plus state law enforcement officers. The increase won't be felt immediately because of the training cycle required for an air marshal.

A huge new bureaucracy and new work rules for the massive increase in armed guards will be created, including hours a day such guards can work and how to schedule them efficiently.

○ Standards for passenger screeners. Congress has passed legislation mandating that 28,000 passenger screeners become federal employees by the fall of 2002. Pay for screeners will go from $15,000 to $35,000 annually. The federal takeover will cost about $2.5 billion a year and will be paid for with a passenger fee of $2.50 to $5.00 per flight.

It's widely agreed that screeners need more training.

The most favored option to ensure that screeners effectively carry out their responsibilities is for them to become federal employees, with the federal government setting the employment and training standards, including pay. The government will also supervise and test the screeners to be sure there is standardized compliance throughout the country. Tough discipline will be imposed on all screeners should they fail to detect an item of contraband.

Screeners will be required to undergo a 10-year background check, the same as anyone else with airport access, and no one will be hired with a criminal record. Screeners will also have to be U.S. citizens.

○ Positive baggage match. Currently, checked baggage of passengers on international flights is checked against the passengers list for a positive baggage match. If the baggage is loaded in the belly of the airplane and the passenger isn't in his or her assigned seat at departure time, then that passenger's bags are taken off the aircraft and left behind. Today there are computerized systems that tell baggage loaders exactly where in the belly the bags are loaded so they can be quickly taken off the aircraft with little delay. Such baggage match will be adopted for domestic baggage as well in the near future.

○ Cockpit security. Airlines have reinforced cockpit doors with bars that limit forced entry into the flight deck of an airliner. Cockpit doors had been designed to be broken into with 130 lb of pressure but have been redesigned to provide a resistance of at least 1500 lb.

The government is requiring all cockpit doors to be replaced with sturdy new doors that are impregnable by the fall of 2002.

O Carry-on luggage. To have quicker yet more thorough screening the government has set limits of one carry-on bag per passenger plus a purse, briefcase, or small portable computer. Any additional luggage will be required to be checked in the aircraft belly.

Knives of any kind, baseball bats, pool cues, ski poles, and, of course, guns are prohibited carry-on luggage.

There are no restrictions on nail clippers, safety razors, or tweezers or on syringes as long as documented proof of medical need is provided.

O Cargo screening. Equipment that can effectively screen checked baggage is on the way. Some are expected to screen 1500 bags per hour which makes it practicable for every bag to be screened prior to being loaded onto the aircraft.

The new equipment will also have the capability of three-dimensional screening similar to a CAT scan. Currently only 5 percent of all baggage is screened. By 2003 all checked baggage will be screened by these new devices or similar ones for hidden bombs, dangerous goods, or explosive materials.

The acceptance of cargo on passenger airlines has been restricted to letters certified by the post office to weigh less than 16 oz. Cargo from unknown shippers or unknown goods in the packages are restricted from being flown on airliners.

○ Armed pilots. The new law allows pilots to carry guns subject to approval by the airline they work for.

○ Foreign airline standards. Standards of safety being adopted for U.S. airlines will eventually be requested of all foreign airlines flying into U.S. air space to protect U.S. citizens.

Security Costs

Security costs per flight are expected to triple or quadruple per passenger. There are approximately 700 million passengers flying annually in the United States and security costs now are just a little over $1 per passenger. New measures being instituted are expected to raise that cost to between $2.50 and $5.00 per passenger per trip.

Immigration Fast Lane

To assist frequent international travelers speed through immigration check points, companies are developing devices that record the iris of the eye or a hand print. These will serve as individual identification sources just like fingerprints. If passengers pay a fee of less than $100 a year and take a second to have their eye photographed, the computer could identify them instantly so they can zip through immigration without waiting in line.

Of course to obtain this identification, these travelers would have to give the government sensitive personal identification data, which many people oppose doing. A test of this program is about to begin at Schiphol Airport in Amsterdam and at the Frankfurt Airport in Germany.

Profiling

America has become sensitive over the years about personal privacy. How the benefits for safety of the many versus the concerns of the few come into focus in the near future will determine just how thorough safety checks are. Profiling will become a standard technique of ensuring passenger safety.

Chapter 4

What the Airlines Are Doing to Protect You

Cockpit Security

To ADAPT TO NEW SECURITY NEEDS, AIRLINES have placed strong steel bars across the cockpit side of access doors. The FAA has given emergency design approval for such bars. Such changes would generally take up to a year. However, although there are over 7000 airliners in the United States it took only about 30 days to make such emergency installations on all U.S. aircraft. That was unbelievably fast.

The best defense an airline can offer against a hijacking while an aircraft is in the air is an impregnable cockpit door. The cockpit doors, which were initially designed as vital breakaway doors for immediate access to the crew in the cockpit in case of an emergency, are being redesigned.

The two major airliner manufacturers—Boeing and Airbus—plus other companies are rushing permanent designs for doors that can withstand bullets, bullies, and bumping. The doors will be installed so hinges can't be broken off with anything less than a strong crowbar, which would take time.

There was talk of confiscating from flight attendants keys to the cockpit door. But all door locks for Boeing aircraft worldwide use the same key and thousands have been issued over the years making that solution impractical. Instead new special tumblers with different keys are being installed in all cockpit doors. Only pilots will have keys—not flight attendants.

Other techniques are being considered as well. At El Al the door can only be operated from the inside by a pilot pressing a button, which controls an electronic lock release. So it doesn't matter whether anyone has a key or not. No key will work in flight. Similar ideas are being considered by some U.S. airlines.

Pilot and Crew Training

Rules airlines lived by since the first hijackings are obsolete in the twenty-first century. The idea of not confronting hijackers but instead making an effort to win their confidence to get the aircraft safely on the ground no long applies.

U.S. pilots will need to be mentally reconditioned in how they respond to any hijacker threat against a crewmember or passenger. If word is communicated to the pilot in the cockpit, probably through an interphone, that the hijackers are about to kill a flight attendant or passenger unless he lets them have access to the cockpit, he'll just have to bite his lip and respond, "Go ahead, that's okay with me, but you're not getting in the cockpit."

That's exactly the mindset and attitude of every El Al pilot and one of the contributing reasons that the airline hasn't had a hijacking since 1968 in spite of being a high-threat target.

Some airlines are also experimenting with self-defense training, especially for flight attendants.

There are some simple self defense techniques that can be learned without a person's becoming a karate black belt.

Carry-Ons

There are some arguments about limiting all carry-ons to a single purse or briefcase in an effort to ease the workload at security checkpoints and to prevent smuggling of weapons. This limitation has several drawbacks: Most passengers want some carry-on with them so that they can do work on long trips. They want to supervise valuables like cameras, jewelry, computers, and other such items not covered by airline insurance if checked. In addition, lack of belly space in some airliners to handle all the carry-on baggage and the potential of theft by baggage handlers and breakage of fragile items is another concern. These concerns are probably the realistic reasons the FAA still allows one purse or briefcase and one other piece of carry-on luggage that will fit into the overhead rack or under your seat. It seems to be a workable compromise.

Preflight Checks

Airlines are also doing daily cabin searches for contraband or anything that could be smuggled aboard the aircraft and used as a weapon.

An additional ID check prior to boarding is being done by most airlines as well as a secondary random carry-on bag search. Prior to push back from the gate most airlines have done or do an instant FBI computerized check of passenger names.

No one is allowed on board whose name appears on suspected terrorist lists provided by the FBI. Currently there are more than 1000 names on the list.

The FBI has relaxed old rules about sharing such information with the airlines. There is now a cooperative spirit.

All metal eating utensils in airport restaurants have been replaced by plastic utensils.

In-Flight Procedures

Some airlines are prohibiting passengers from forming lines to wait for the forward lavatories. Passengers must either go for the rear bathroom or wait in their seats until there is no line.

The idea is to limit traffic around the cockpit area. Some individual pilots and flight attendants have even taken it upon themselves to prohibit passengers from using the front bathrooms during flights.

The policy for flights at Reagan National Airport is even tougher. Passengers are prohibited from leaving their seats for 30 minutes after takeoff or for 30 minutes before landing at the Washington, D.C. airport.

Airlines are leaving curtains or dividers between cabin classes open to allow for unobstructed view by all in the aircraft.

Seat belt signs are being strictly enforced.

Cabin crews are working more closely together than ever to facilitate immediate reporting of suspicious activities to other crewmembers.

Any preflight beverage service during the boarding process is being suspended to allow flight attendants to focus on passenger boarding.

Airlines are analyzing how to prevent deactivating of emergency radio and radar signals in the cockpit that sound an alarm of a hijacking to a ground station.

All airlines are urging passengers to keep their seat belts fastened throughout the flight. One reason is that any final desperate aircraft maneuvering effort by pilots to throw hijackers off balance to prevent a hijacking could, of course, be fatal to anyone standing or sitting without a seat belt fastened.

There is some talk of installing cameras that will continuously video the cabin, with a monitor in the cockpit. But where to place the monitor in the cockpit and how to string the necessary wiring are complicating factors.

Cargo Limitations

No cargo or mail is being transported except:

1. Human remains, organs, and blood and tissue, in accordance with specific procedure spelled out by the FAA.

2. U.S. mail, which may only be transported after obtaining the following
   written certification from the tendering official: "The Postal Service certi-
   fies that no mail parcels sixteen
   ounces and over are being tendered to
   the accepting passenger air carrier."

3. Official U.S. diplomatic mail and
   pouches.

4. Material classified in the interest of
   national security by the U.S. government.

5. Cargo received from known shippers.

6. Cargo received from FAA-approved
   indirect Air Carriers not identified as
   originating from an unknown shipper.

7. Cargo received from another air carrier subject to security directives or a
   foreign air carrier with a security
   program approved by the FAA.

Profiling System

The new profiling system builds on a little-
publicized system that has been widely
employed by the nation's airlines for
the past four years. It is known as
the Computer-Assisted Passenger Pre-
screening System, or CAPPS. Run by the
FAA, CAPPS was created after the midair
explosion of TWA Flight 800 over Long
Island Sound in 1996, which was ultimately ruled an accident. CAPPS uses

basic data disclosed by travelers when they reserve and buy tickets—such as their names, addresses, and how and when they paid—to look for patterns that could point to terrorism.

CHAPTER 5

What You Can Do to Protect Yourself

THE GOVERNMENT IS SETTING UP REGULATIONS to protect passengers, airlines are taking actions to safeguard flights, and airports are instituting new security procedures, all to thwart any terrorist threat.

Passengers can also take steps that will add an additional layer of insurance for their personal safety. Actions you can take include:

○ Check departure time one hour before leaving for the airport to be sure your flight is on time.

○ Ensure you leave a complete itinerary behind including flights, hotels, and contact numbers, one for your family and one for your office.

○ For domestic flights curbside check-in may be best. E-Tickets make this fast. Most airlines will allow those with e-tickets with faxed confirmations to go to the airline club or directly to the gate as long as you have a receipt or printed confirmation. The airline clubs, especially on the secure side of the terminal, are great inconspicuous places to hang out while waiting for your flight. This means you can go directly to security check point lines without having to go to the ticket counter, as long as you did curbside luggage check or have only carry-on baggage.

○ To avoid bag searches and extra security attention don't buy a one-way ticket, don't change your itinerary less than 72 hours before your flight, and don't pay cash, especially for an expensive ticket.

○ Aisle seats are best for men, and center and window seats best for women and children. If there is trouble, you want men to be able to get to the action fast.

○ Whether you board the plane first or last, keep an eye on your fellow passengers and do your own screening and profiling. If you feel

uncomfortable with any passenger
let the gate agent or flight attendant
know. If they remain on the aircraft,
and you still feel uncomfortable don't
be afraid to ask the airline to
reschedule you on another flight.

○ Wear your seatbelt at all times, but
be sure that you know how to
unfasten it in a hurry. If you have to
take any self-defensive action, you'll
want to be able to move quickly.

○ Do not wear chinos. This may
identify you as an American.

○ Get to know your seatmate—
especially if you sit in first or
business class. In a terrorist hijacking
you and that person may have to
work together as the first line of
defense.

○ Should you be victim to a hijack
don't wait for an air marshal. At this
stage there are very few trained air
marshals on the 30,000 plus daily
flights in the United States. It's like
the Wild West—you're deputized to
defend yourself and others.
Remember the chances are they'll be
more than one or two hijackers, so
you may want to yell for help. You've
got to fight back. You can use shoes,
full soda or beer cans, pillows, and
blankets to attack the hijackers. You
won't be alone—not in this day and
age. Every hijacker must be treated

as if he is on a suicide mission. Smother the person with as many bodies as possible. Put blankets over his head so he can't see. Yell for help from fellow passengers. Think in your own mind—it's either you or him. Belts and shoelaces make for good handcuffs. Be sure that the flight attendants are keeping the cockpit aware of what is going on.

Chapter 6

Guns in the Cockpit

As we enter the new era of terrorism, where hijackers use airliners as powerful bombs, the cry for guns in the cockpit for pilots has captured the attention of the press and some members of Congress.

The president of the Air Line Pilot's Association (ALPA), the largest pilot union, has testified to Congress that the union wants pilots to carry hand guns in the cockpit. Some argue that the pilots are entrusted with equipment worth tens of millions of dollars so why can't they be trusted with a gun? Of course, they aren't trusted with that aircraft without years of training and experience. Would the same be true for a gun?

This leads to the argument of what is the best way to protect an aircraft and its passengers.

Some would say guns in the cockpit is a knee-jerk reaction. There is no such thing as a "friendly gun" on board an aircraft. There is no such thing as a "friendly bullet" when it penetrates a very vital avionics panel that is concealed beneath the floor. So careful thought must be given to the idea.

Many practical issues have to be considered: How does the pilot get the gun aboard the airplane? Does he carry it from home? Does the airline check it out to him and collect it when he gets back from a trip? If he has a layover what does he do with the gun? What type and caliber gun should be used? What type of bullets? How does the pilot clear security?

Could a terrorist get outfitted in an easy-to-purchase pilot uniform with fake or stolen ID and walk right through the security checkpoint and into the secure area with a gun? Could he then go to the men's room and change into some other clothing? If a pilot is carrying his weapon could someone on the secure side of the airport mug him in a bathroom and take his gun?

Should there be a safe on the aircraft for the gun? In most cockpit seats the pilot is completely surrounded with the instruments and controls necessary to operate the aircraft. The result is that there is no room for a safe within arm's reach. Where should the safe be located? Who should have access to the safe? How do you control access so that a caterer or a mechanic or ground service person with access to the aircraft on the ground before the crew arrives doesn't steal the gun? What if the pilot says he put the gun in the safe, but the next crewmember says it's not there and neither story can be proven? Who is liable? What happens if that gun is used in another crime? Who is liable?

Where should the gun be placed during the flight for quick access? Should the pilot wear a holster? Should it be placed in a side pocket? Should both captain and copilot have guns? What if the terrorist gains entry with a flight attendant in front of him as hostage and shield? Does he shoot through the flight attendant to try to kill the terrorist? What happens if several suicide-bent terrorists rush the cockpit and the pilot shoots one or two but the others get his gun and use it on him when they don't have one themselves?

What if the pilot makes an awful mistake thinking there is a terrorist attack when there isn't one and innocent people die?

Does a pilot need firearm's training? If so where does he get it? How long does suitable training take? Who pays for the training? The Government? The cash-starved airline? Who fills in for the pilot while he is being trained and not flying his usual flights? What are the legal requirements for carrying the gun?

There is an argument that most airline pilots are former military pilots. Years ago that was true but today since the end of the draft most commercial airline pilots came up through the ranks as student pilots, then instructor pilots, then commuter pilots, and then finally into the cockpit of a large jet airliner.

Those airline pilots who claim military flying background mostly flew transport aircraft and had nothing to do with firing guns. Those who actually flew fighters shot their cannons or missiles at distant targets. Military pilots weren't trained for a phone booth type engagement. That is not to say a lot of military guys don't hunt or shoot guns, but they are in the small minority.

These are the kinds of questions with no answers that make airline management nervous about the thought of guns in the cockpit. If airline management doesn't want a gun in the cockpit and an innocent person gets killed by accident in a nonhijack situation, does the airline have liability responsibility for the pilot's actions?

These are some of the realistic everyday questions that need to be thought out and answered before the Congress or the FAA arbitrarily says it's okay for pilots to be armed with guns. We've got to be sure what we're doing before any weapon is allowed aboard the aircraft.

So what's the solution?

It is generally agreed that barred doors and the installation of some sort of strong bulletproof Kevlar door, highly trained and skilled armed air marshals aboard the aircraft, plus effective screening and profiling as passengers file through the terminal gates are the best preventative measures against terrorist attack.

Of course, the problem is that the terrorists have proved innovative and creative in their efforts and have rarely used the same techniques twice in a row. Therefore a lot of thought needs to be given about how to use all means available to thwart all imaginable terrorist acts.

CHAPTER 7

Airport Security

AIRPORTS HAVE PROVED TO BE HIGH-RISK TER-
rorist areas.

Most European and Mideast airports have
crowded facilities. They just aren't big
enough to handle the volume of traffic that
exists today. That means people are
jammed together, making airports inviting
targets for terrorist intent upon killing and
maiming indiscriminately.

Security has been beefed up domestically
and internationally. Screeners now are
painstakingly thorough as slow lines snake
through airport lobbies. Every entrance is
heavily patrolled, and passengers are

checked with handheld detectors. Passengers may be frisked and searched.

Nevertheless, at all airports there are important precautions every passenger should take for self-protection.

Say goodbye to friends and relatives in the car when they drop you off at the airport. Avoid exposing them to a crowded airport and high-security situation.

Be sure to check departure times before leaving your hotel or home. Is the flight on schedule? Avoid spending unexpected hours at the airport waiting and exposed to dangers. But be sure you allow ample time to clear the screening checkpoint.

Proceed alone or with your traveling companions through the check-in process and security checks, without any nonflying guests or hosts.

Do not take a fancy limousine to or from the airport. It just calls attention to you as someone special or rich. Use an ordinary cab.

Stay away from any people who are using such vehicles or anyone receiving obvious special attention from the airlines. Fade into the background and keep away from other passengers as much as possible.

Be sure you lock all your luggage—not just to avoid theft, but to discourage anyone from putting anything in it for shipment aboard an airliner.

Curbside baggage check-in for international flights in the United States and abroad is generally prohibited. The idea of checking your bags at the counter ensures that only luggage of a ticketed passenger gets aboard the airplane. Airlines are now adopting systems to ensure that all luggage checked belongs to passengers who are on board the aircraft. This is to ensure that a bomb is not aboard in a stray piece of luggage.

All major airports now employ duplicate checks of baggage and travel papers to ensure that if a terrorist evades one security search, there will be others.

Some airlines have duplicated the airport's security systems with their own as well.

Some carriers on domestic routes make computer checks of passengers and their luggage to ensure that the plane does not carry the luggage unless the passenger sits in his or her assigned seat. This is true on all international flights. Other carriers require passengers to point out their bags for loading on the plane, just prior to the passenger walking into the airplane.

If someone should ask you to carry a package for them to your destination, refuse, unless you personally know the individual as a relative or friend and you yourself have inspected the contents of the package.

If you are asked to transport such a package at an airport by a stranger, say no and then quietly notify airport security officials who may ask you to help identify the person.

There are two thoughts about check-in. Arrive very early, long before the crowds, check in at the uncrowded ticket counter and proceed through security and immigration, and then wait in the passenger boarding area, which should be the most secure area.

The other thought is to arrive with just enough time to be processed at the ticket counter, security, passport control and to walk to the gate just prior to departure. This eliminates waiting time at the airport and thereby lessens the chances of being involved in any problems.

The idea is to minimize the time spent during check-in procedures at the counters and to avoid crowds, which are inviting conditions for terrorists.

Only one person in a small family group should take care of the check-in procedures, although you will need to show picture ID cards to match the name on each ticket. The remainder of your party should stay in an area remote from the crowds but within easy access to the check-in agent. If you are with a large group, including student groups, each member will be required to check in and account for his or her baggage individually.

In the terminal area and at the departure gates of extremely sensitive cities, do not stand or sit next to large plateglass windows or walls. If there is nowhere else to be, then keep your back to the glass. Wear as heavy a coat or clothing as you can stand for protection from shattered glass or debris.

It is ideal to sit or stand next to a column or post that is a supportive structure for the building.

Think about what you'd use as shelter if there was an attack. Is there a vending machine for soft drinks nearby? It could be used for protective cover. Are concrete barriers nearby? What about sofas or chairs that have plenty of padding? What else can you spot that could insulate you from debris, shrapnel, or bullets?

Among the safe areas to wait are airline members-only clubs or lounges. They vary from airport to airport as to whether they are located before or after security. But they are isolated from the general open areas of airports. Usually such clubs are barely marked, and they have solid doors and walls which make it impossible for terrorists to see so many people congregated together. These clubs are remote from terrorist attack and are a good safe place to wait for your flight.

While waiting, look around and become familiar with all exits, including emergency exits. Don't just concentrate on knowing about exits that go out into the street or those in front of the terminal, but look for exits in other directions as well.

Stay away from stray baggage left unattended. If you see any luggage or boxes sitting alone, immediately notify airport security.

Telephone booths (although most have been eliminated in more modern airports), trash containers, and enclosures of all types are excellent places to hide bombs. Many airports have trash containers with clear vinyl liners and holes in the supporting container that allow you to see the contents in the trash. If you see large packages in those waste containers, notify airport security.

Once you have completed the check-in process at the ticket counter, proceed immediately through the security check and get into the secure or protected area of the terminal. Don't stroll or lounge around in the open public areas of the building near the unprotected entrances.

Terrorists will go for easy access areas that are crowded. Either do your shopping in the city or behind the security barriers inside the airport terminal.

Should there be an evacuation alert of the air terminal, stay in the center of any crowd. Don't panic and race out in front or drag too far behind. We're talking survival— and a crowd around you is like secret service agents surrounding the President; it provides a buffer between you and danger.

If there is a commotion or problem anywhere, quietly get out fast, but don't run or you'll be suspected. There's nothing you can do, so don't let your curiosity get you involved. There could be secondary attacks. Keep a low profile.

While waiting, keep alert to any indication of trouble. If there is a sudden concentration of security guards in your area, be sensitive to what is going on.

Once aboard the aircraft, check under your seat or in the overhead for any stray packages. If there are any that don't belong to you or your seatmates, notify the flight attendant immediately.

Upon arrival at your destination, don't be in too much of a hurry to claim your luggage. Luggage arrival is often slow. Lay back; this is an easy-access public area. Stay away from the crowd and wait until it dissipates. It will cost you only a few minutes, but provide you with enormous protection.

Once you've left the airport area, peel or scrape off your luggage security clearance tags and destroy them, so that a terrorist can't use them at a future time.

Pay attention to your intuitive feelings. They are usually right. But don't let unjustified fear precondition those feelings.

Much of this may sound like a bunker mentality, but it's only wise to develop and follow actions to protect yourself from potential danger

It bears repeating that your chances ·of being involved in an aerial hijacking are minuscule, but why not be alert and take simple protective actions that could make a big difference if there is trouble?

If you travel frequently, special precautions will become second nature to you and will not interfere in the enjoyment of your traveling adventure.

Some of the best protection possible comes from alert passengers who share any potential problems with airport security. Don't depend upon airport security alone. You are all in the terminal together, and security officers will appreciate any help your extra set of eyes can provide. Don't hesitate or worry about what the security people will think of you. They're interested in anything you have to say. They won't ridicule you or make jokes about your information.

CHAPTER 8

Preparing for Your International Trip

IF YOU PLAN AHEAD, YOU SAVE MONEY. IF YOU don't, it costs you money.

The same could be said about your life when planning a trip, especially to a high-threat area. Begin your preparation for your overseas trip well in advance.

Although the odds are remote of ever being involved in a terrorist hijacking or attack, there are definite preparations to make in organizing your trip that could go a long way toward ensuring a safe and pleasant journey.

Time spent considering the possible consequences of a terrorist attack could clarify your planning. Numerous reports of those who have been held hostage have confirmed this.

These thoughts and ideas are not meant to scare you, but just to provide some simple guidelines. They are offered as a start to get you thinking about the most effective way to protect yourself in international travel.

1. Check your calendar. We have holidays and special events that are significant to us. So does the rest of the world. Are you traveling on one of those days? If it is not necessary to travel at such sensitive times, why do it?

   Match up your schedule with significant political events of terrorist countries, and avoid travel on those days to that area of the world.

2. Check your passport. How much travel does your passport show? Is your occupation listed on it? Does it have Israel stamped on it? If so, get a new one. It's probably best not to put information on your passport that could help your captors know too much about you or your relatives.

   Government officials and military should travel on a normal blue tourist passport in high-threat areas.

3. Fly a nonstop flight to your destination if possible. The fewer number of stops, the fewer chances of a hijack. The idea is to leave the United States

and reach your destination nonstop. Not all flights overseas are the same, so check them out carefully.

4. Consider the type of aircraft. The smaller, narrow-body jet airliners like the Boeing 727 or 737 are the traditional hijackers' favorite targets. They are smaller planes, which makes it easier for fewer hijackers to control the passengers.

The hijackers seem to avoid the large wide-bodies, such as Boeing 747s, 777s, or 767s; McDonnell Douglas DC-10s, Airbus A-330s, and 340s; and Lockheed L-1011s. None had been hijacked overseas since 1970, until the Pan Am 747 Flight 73 was hijacked in Karachi, Pakistan, in September 1986.

The problem with hijacking a large wide-bodied airliner is that there are just too many people to control for the small number of hijackers usually on board. The more hijackers who try to get aboard, the greater their risk of being detected. So they have been sticking to the smaller planes.

A narrow-body jet generally has more than 100 passengers, which gives the hijackers enough hostages to get the media attention they crave.

However, if the hijackers are intent on killing people or executing a suicide mission, the larger jets make a more spectacular event.

5. Consider flying at night. The hijacking pattern has always been to intercept flights in daylight. The hijackers want to see what they are getting. They want to see where they are going. They seem to feel they have better control of events at the start of their terrorism in daylight. Think about scheduling night flights when you are in high-risk areas.

6. There has been discussion about which airlines to fly—U.S. carriers or foreign carriers. It's generally considered a trade-off. Some foreign airlines like Sabena and SAS are from neutral countries and the thinking is that terrorists won't bother them.

   On the other hand, if target groups like Americans, Israelis, French, and British pack those planes, it won't make much difference to the terrorists whose flag is on the tail. It's not the airline they're after, but the innocent passengers on board who serve their purposes.

   Security on U.S. carriers and El Al is considered superior to any other of the world's airlines. Lufthansa has also instituted crack security.

   There has been a renewed interest in charter flights of both large and small jets. Many of the charter airlines are seeing a rise in their business because they operate without all the

fanfare of the large scheduled airlines and are lesser known.

7. Get information on the places you
   are going. What's happening? What's
   going on now? Have your travel
   agent check for any U.S. travel advisories for the areas you are intending
   to visit. Study them so that you
   undersatnd what effect they could
   have on your trip.

8. First-class seating calls attention to
   yourself. You may want to evaluate
   flying first class versus flying a little
   more uncomfortably, yet anony-
   mously, in business or coach class, in
   a high-risk area.

   It's in the first-class compartment
   where the terrorists usually set up
   their command post—right behind
   the cockpit. If they are going to control the plane, the terrorists need to
   control the cockpit. The main exits
   are located there, too, and that helps
   the terrorists control the situation.

9. Wear no identifying items. Leave at
   home all class rings, college rings,
   military academy or other military
   rings, as well as all jewelry with any
   Hebrew language or Jewish emblems
   or markings. If you absolutely need
   them at your destination, put them in
   your checked luggage so they won't
   be with you in the passenger compartment. But remember, maximum
   insurance on luggage is $2,500 per

bag with some kind of proof that you have items of that much value in your bag.

The insurance doesn't cover jewelry, cameras, computers, or cash. Before checking such items, check the back of your ticket for such restrictions.

10. Leave all fancy jewelry at home. It's not necessary for travel, and large diamonds, rare stones, or flashy gold will only call attention to you. This includes rings, necklaces, and watches.

In the 1985 TWA Flight 847 incident, the hijackers were common thieves as well. They took all the passengers' money and jewelry. It was estimated their haul was a quarter of a million dollars. Why open yourself to the risk of special attention and loss of valuables?

11. Before leaving home, clean out your purse or wallet. Strip it of everything except what you actually need for the trip. Take nothing else.

Probably all you'll need for the trip besides your passport are a couple of credit cards in your own name—not company name—some cash and mostly traveler's checks. Be sure you have some identification like a driver's license that has your home address. But you won't need military retirement cards, military ID cards, or government documents.

There's an old adage, "while traveling take half as many clothes as you think you need and twice as much money."

If you're military, travel on a regular blue passport. Ship your military ID card in your luggage. Leave behind all dependent ID cards or ship them in baggage. You don't need political party, veteran, or any controversial membership cards or police badges.

Don't take your checkbook—it will reveal your wealth. But you could rip out some blank checks and keep a tally of what you spend as you travel.

Be sure each member of your party is self-sufficient with traveler's checks and credit cards, in case you become separated.

12. If you are on a business trip, put business cards, company letterhead, personal tax statements, and all iden-tifying material with your checked luggage. That should include sales brochures, financial statements, and annual reports.

Files with letters of sensitive data and any information that could cause problems if a terrorist reads them should be in your luggage.

Be especially careful if you're involved in the defense industry.

Business people are now sought after, so why take the chance of needlessly calling attention to yourself, especially if you are part of some high-visibility multinational corporation.

13. Carry a card or something with your blood type and any important medical information in case of emergency.

14. Have simple luggage tags with name and address, including country—but no business name or address or attention-getting designs, especially those that would include national flags. These tags should be on all your hand-carry baggage as well as checked luggage.

15. If you are taking medicine, be sure you carry several days or maybe a week's extra supply on you. If something should happen, you may not have access to your luggage. Although most captors will provide you with medical care to ensure your survival, there could be a delay before medicine or a doctor is available.

16. If you are traveling with small children, be sure you take in your carry-on luggage extra diapers, milk and food, plus any medication the child may need. A bottle of liquid baby aspirin is always good to relax children and help them sleep during a time of stress.

17. When purchasing your ticket, try to get as many seat reservations as possible in advance.

The ideal seat, if you are in good physical condition, is the aisle seat near the over-the-wing exits. Second choice would be near door exits. Infirm or handicapped people will be assigned specific seats by the airline and, along with children, are not allowed in a seat next to a window exit. The reason is it is important to have an able-bodied person in that seat.

On TWA 847 there was pistol-whipping by hijackers of men sitting on the aisles. To avoid that, men could sit by the windows. The hijackers seem to have dealt toward women with somewhat more respect. Women seem less of a threat to them.

However, a man in an aisle seat is in a better position to try to attack any hijacker intent on turning the aircraft into a missile aimed at some ground target.

18. Plan your wardrobe to avoid clothes that will set you apart from the crowd. Is that expensive fur necessary? Will the bright-colored dress, loud jacket, or big hat set you apart from your fellow travelers?

19. Arabs are offended by magazines like *Playboy*, *Hustler*, *Penthouse*, and the like. Don't get caught with one.

20. Try to learn a little of the local language of your destination to help you in an emergency.

21. There should be a brief family discussion about what to do in case anyone in the family is taken hostage. It should include who would be the point of contact and where family members would go at such a time.

22. Leave behind a list of regular payments for recurring expenses like home mortgage payments, insurance payments, and taxes.

23. Establish a joint checking account so that if separated, either spouse could write checks.

24. Update your will and ensure that your executor and lawyer have a copy. The original should be placed in your safety deposit box. It should contain a designation of a guardian for your children in case something should happen.

25. If you don't have a safety deposit box, you can obtain one for a small charge at most banks or savings and loans. You should place in it all important records and papers, including trust deeds, insurance papers, mortgages, tax records, and names of doctors and dentists should medical records be necessary.

26. You may want to consider executing a power of attorney for each adult member of your group. It could be exercised in case there are any emergency problems while you are gone

and there is need to transact business on your behalf.

Should you become a captive, you'll be happy you've taken these steps of preparation. If you procrastinate, put off, or delay making the proper decisions to organize your trip, you could have feelings of guilt as you sit in captivity wondering why you didn't do what was necessary before you left. Don't put off until later things that need to be done now to get ready for your trip.

CHAPTER 9

Overseas Traveling Tips

PEOPLE TRAVEL FOR A VARIETY OF REASONS: tourism, business, diplomatic and governmental activities, military purposes, and newsgathering, to name a few.

Travel Advisories

The U.S. State Department issues travel advisories concerning travel to every country in the world.

The travel advisories alert Americans of dangerous or life-threatening conditions, and they list countries with terrorist problems or unstable governments. They warn Americans to avoid certain countries or to exercise travel cautions. Those countries the State Department currently warns Americans to avoid (as of November 2001) and the dates of the travel advisory issue are:

Avoid	Travel Advisory Issued
Macedonia	10/22/01
Sudan	10/05/01
Indonesia	09/27/01
Tajikistan	09/26/01
Pakistan	09/25/01
Kyrgyz Republic	09/21/01
Turkmenistan	09/19/01
Yemen	09/19/01
Iran	08/24/01
Sierra Leone	08/20/01
Israel, the West Bank, and Gaza	08/10/01
Sri Lanka	07/24/01
Iraq	07/20/01
Libya	06/06/01
Liberia	05/31/01
Algeria	05/31/01
Central African Republic	05/30/01
Solomon Islands	05/01/01
Guinea-Bissau	04/30/01
Colombia	04/17/01
Bosnia and Herzegovina	04/13/01
Democratic Republic of Congo	04/11/01
Somalia	02/16/01
Federal Republic of Congo	04/11/01
Federal Republic of Yugoslavia	02/13/01
Afghanistan	12/12/00
Burundi	12/07/00
Angola	09/08/00
Lebanon	08/28/00
Albania	06/12/00
Nigeria	04/07/00

Countries for which the Department of State has issued travel warnings to be cautious about are listed in the accompanying table.

Exercise Caution	Issued	Expires
Bangladesh	10/26/01	01/09/02
Worldwide Caution	10/23/01	04/19/02
East Timor	10/22/01	03/31/02
Philippines	10/04/01	11/03/01
Luxembourg	10/04/01	11/03/01
Italy	10/02/01	01/06/02
Niger	09/28/01	01/06/02
Georgia	09/26/01	01/08/02
Uzbekistan	09/22/01	01/08/02
Rwanda	09/19/01	03/18/02
Uganda	09/07/01	12/06/01
Colombia	09/07/01	12/03/01
Middle East Update	08/10/01	11/10/01
Philippines	06/26/01	11/01/01
Malaysia	06/13/01	12/06/01
China	04/19/01	12/27/01

Of course, these problem areas are subject to change and more could be added to the list or some of those listed deleted. For complete information as to the status of these countries listed go to the U.S. State Department Website for travel information at *www.travel.state.gov.*

If you are planning a trip and have doubts about safety, you can contact the American Citizens Emergency Center at the U.S. Department of State for automated voice travel advisories by calling 202-647-5225.

To talk with a live person about travel advisories in the U.S. call 888-407-4747 and tell them what countries you are interested in gaining facts about.

If you are overseas and need information or have an emergency, call the U.S. Department of State at 1-317-472-2328.

Your local passport office should have the same information, as well as most travel agencies. The information is also available by mail from the State Department. Write to the Citizens Emergency Center, Office 4811, 2201 C Street, N.W., Washington, DC 20520.

Military personnel should travel in civilian dress. If your hair is cut in the short, distinctive military manner, it would be wise to hide it and disguise yourself by wearing a nondescript baseball cap. As a member of the military, you represent a lot more about America to the terrorist than does a normal tourist.

Emergency first aid training is always handy to have, especially if you are traveling to a remote part of the world. It becomes even more valuable in a life-threatening situation.

Upon arrival in a foreign country, it's wise to check in with the American embassy. Every embassy has a security officer. When you check in, the officer will be happy to brief you on what's going on in the country, the risks, what to avoid, the political situation, and what places to visit first. Or you could write the desk at the State Department that specializes in your destination prior to your departure for any information they may be making available. Most embassies provide handouts of what to do and not to do and places to avoid.

Safety Tips

It's smart to avoid sidewalk cafes frequented by other Americans, including tourists and the military. Mix in with the local people.

When you are in public places, keep a low profile. Don't do anything to call attention to yourself—especially as an American.

Read the local newspapers and keep informed of what is going on while you're in that country.

Yell "fire" or "accident" in the local language if you are attacked. Yelling "help" usually doesn't gain as much attention.

Here are some suggestions and cautions to observe when visiting or staying at a hotel:

1. Book a hotel with good security. Although there is no printed guide for a travel agent to refer to, he or she may have enough experience in the area to be of help.

2. Avoid the top tourists hotels. They are easy targets and are symbols of

a variety of things to terrorists, which could make them and you a target.

3. Don't share with hotel employees what your company does or why you are there.

4. When answering your telephone, just say "hello"—don't give your company name or any information.

5. Don't accept deliveries you don't expect.

6. Although there have been no incidents so far, if you're part of an organization holding a meeting in the hotel, it would be wise not to have any welcome signs in the hotel lobby. Registration tables should be in the meeting room. You should only wear your convention name badge in the meeting room and not around the hotel or in the streets.

7. Most fire trucks have ladders that will reach only as high as the sixth floor. On the first or second floor there is the threat of ordinary crime and break-ins. So generally the third through sixth floors are the safest areas of a hotel.

8. Don't go to the window if you hear gunfire on the street or a bomb alarm goes off.

9. Don't work in front of large windows.

Car bombings are the most prevalent single threat by terrorists. If you use a vehicle in a foreign country, observe the following cautions:

1. Rent or use an ordinary vehicle made in the country that you are visiting. Nothing flashy or showy or American. A Mercedes should only be used in Germany.

2. Use a car with a hood latch on the inside, not the outside. Be sure the hood has not been tampered with and there are no unnecessary wires dangling around.

3. When you rent a car, don't leave anything visible showing through the window that identifies you or your nationality.

4. Lock the car when it is parked.

5. While driving, watch where you are going. Watch out for barricades.

6. In more remote countries, like China, use taxis or limos. Don't drive yourself. Should you have an accident in a place where your insurance is meaningless, you could land in jail for an indefinite time.

7. Don't get blocked in when you're driving. If traffic up ahead is all blocked up—it could be by design. Leave room between you and the car in front to do a quick U-turn. Don't get

trapped or fenced in. You must have room to maneuver or pull away in an emergency.

Although these are all good tips, the most important consideration in foreign travel is alertness for any problems and thinking ahead about what you would do in any given situation.

CHAPTER 10

How to Behave in a "Traditional" Hijacking

ALL YOUR BEST EFFORTS HAVE FAILED AND YOU'RE hijacked—you're a prisoner against your will. Now you just want to avoid being selected for special treatment.

You realize this is a "traditional" hijacking for extortion or to cross a border and not a suicide terrorist attack. Although your life is at stake, in reality you are nothing more than a pawn in a much larger chess game of political intrigue. You'll want to do everything possible to survive this upcoming ordeal.

Your first reaction will probably be one of terror, fright, and grave concern for your safety. If you can survive the first 15 minutes, you'll probably survive the entire ordeal. All your advance planning will now come into play.

Your pilots and flight attendants have had training for this event. They are going to take some very special actions and do everything they can to ensure your safety, but they're not going to have a lot of leverage in dealing with the hijackers. Passengers should rely on the advice of flight crews. In recent months, their training has increased eightfold.

There will be instructions right at the beginning from the hijackers. Follow them. Don't try to be a hero. Don't confront, antagonize, or argue with your captors.

The hijackers are also jittery and nervous at this stage. They don't know exactly what to expect. They don't know what hidden defenses exist to thwart their efforts. To take effective control of the aircraft they must be forceful. That makes those first few minutes the most dangerous.

Anything or anyone giving them a problem could prompt a violent reaction in return. You want to avoid anything that could cause you to be selected as someone special for special treatment. Just become invisible. Be as low-key as possible; try to be inconspicuous.

Don't ask questions. The hijackers are going to tell you what they want you to know and that's that. Follow instructions.

The hijackers will probably have all passengers keep their seat belts fastened. They may even have you put your tray tables down to inhibit any movements you might try to make to hinder their activities.

In addition, they may require everyone to pull down the window shade. Being in the darkened aircraft can be confusing and intimidating over a period of time. You may have to exercise some mental discipline to handle it.

They've hijacked you for a purpose—either to make a statement to the world or to hold you hostage in return for something. Don't flaunt your citizenship. Don't carry on like you're a big wheel, have any official connection, or are someone important. You may become more important in this drama than you want to become.

Don't do a lot of complaining about the general situation. Such behavior will only draw attention to you.

If you have a specific life-threatening problem let the hijackers know. Although these are callous and murderous people, the general attitude in the past has been that you're only good to them as a live hostage.

They are going to feel that the passengers are more valuable to them alive than dead. Their purpose is not to kill the whole planeload of passengers.

However, they realize once a single hostage is killed, their lives have become worthless if they cannot escape. So keeping the plane mobile and passengers under control becomes of paramount importance.

Show a quiet, respectful, obedient attitude regardless of the true feelings in your heart. Don't violate the hijackers' orders.

It's best not to talk. If you do, talk in a normal voice. Terrorists are suspicious of whispering.

Don't volunteer to do anything. Especially, don't volunteer to be a spokesperson. You could simply play into their hands of giving validity to their propaganda.

If you are a leader, executive, or person who generally takes charge, this is a time to backpedal and observe.

Don't ask for special permission or favors, such as being able to smoke, to change seats, or to have something to eat.

Only ask to go to the lavatory when you are in extreme distress—not just uncomfortable, but downright miserable. The rear toilets are preferable to use.

If you have any nonalcoholic liquid in your possession when the hijack begins, take a swallow or two. It will have a relaxing effect on the shock and trauma of being kidnapped.

However, don't drink lots of liquids or any alcoholic beverages. If offered an alcoholic beverage, accept it, fool around with it, but don't drink it. You're going to need to be as sharp and alert as possible throughout this ordeal.

AVOID EYE CONTACT WITH THE HIJACK-ERS. Don't stare.

Don't talk to other passengers or attempt to carry on conversations. The hijackers could think you are plotting against them, and the consequences could be severe.

Don't trust anyone on the aircraft you don't personally know with your ideas, opinions or thoughts for action. That person could be an undercover accomplice.

Rest as much as you can. Conserve your strength. Keep alert to what is happening.

Usually an escape effort is most successful when first captured, but it is virtually impossible aboard an airline because it is so confining and the hijackers have all the weapons. Analyze escape ideas with great caution. The chances of success are very slim.

If something should happen and you have a good chance to get away safely—take it. But once you execute your plan, keep moving, don't hesitate or be indecisive.

Don't try to ingratiate yourself to the hijackers. They may take advantage of it, but it could come back to haunt you mentally and probably physically. When you're no longer of use to them, it could work against your best interests. The hijackers will have no loyalty to you once you've given them what they want. However, your fellow passengers will hate you for your actions, which will be perceived to have been taken at their expense. The whipsawing of emotions could be deadly.

Don't confess or admit to anything. Don't offer any political opinions or ideas. After things settle down, the hijackers may try to persuade you that their cause is a just one. Listen, but don't get involved. Explain to them, if you have to answer, you're just not knowledgeable enough to discuss the subject. Don't agree or disagree.

Don't get into any religious discussions. You're dealing with hard-core fanatics; they're not about to be converted to your point of view.

Remember, hijacking you is the biggest event in the life of this terrorist. He has mentally psyched himself up and is willing to give his life for his cause. He may not want to, but he is willing to. Don't put him on the spot where he might have to make that choice—because it will be your life, too.

During your captivity, the hijackers may try to convert you to their views. If you're not informed or knowledgeable on the subject, you might be taken in. Any statements you may make for your captors to publish around the world could show your naivete and what a pawn you are. Or your statements could aid their worldwide propaganda effort—much to your chagrin and embarrassment later. Just listening is the best action.

Let the hijacker settle down and get into a routine of the hijacking and his emotions will settle down, too. He may start valuing his life again a little more.

All your advance planning will now come into play. Having sat in a window seat, you should be out of reach of any pistol-whipping terrorist. If you are on an aisle, keep your feet out of the aisle.

The heavier clothing you wore will be a shield or protection from flying debris or shrapnel from any explosive device that is detonated.

You will not have any identifying jewelry, or special items that could give you strong military, Jewish, corporate, or government identity. You won't have many valuables, so when the terrorists decide to make a collection of all jewelry and money, your loss will be minimal.

Don't attempt to hide or withhold anything. It's not worth the risk. If you have anything that's special or of sentimental value, don't ask to have it exempted. They're not going to make an exception for you. If you can't afford to lose it, don't take it on your trip in the first place.

Accept all food. Eat what is offered, whether you like it or not. Don't wait until later. Keep your energy and strength up. You don't know how long the hijacking is going to last. Most food is basic—just prepared differently.

You'll have the medicines you need, if you planned ahead.

Now is the time to start thinking and mentally preparing yourself for several eventualities. If the hijackers have confiscated your belongings or papers, have some logically thought-out reasons for any items that could be embarrassing to you in this situation. But don't be too long-winded.

Start analyzing the situation and mentally develop plans for what can happen. Check out where you are in relation to the emergency exits and doors. Observe what is going on in the aircraft. Take a mental head count of the number of terrorists.

How many passengers have been singled out for special attention? Why?

What are the terrorists wearing?

What is their race, color, and sex?

What identifying facial features do they have?

Do they have any unusual features, such as a limp, deformed arm, bad acne, or anything out of the ordinary? Are any of them noticeably left-handed.

What are the actual weapons they are using and carrying? Types of guns, grenades, dynamite, rifles? How many?

Focus in on what you are actually seeing. Gather anything of intelligence.

How many men are on the airplane? How many women? How many children? How many infants? You may not be able to count them all, but try—especially if you are a woman. Then if you are released early, this will become valuable information in helping the authorities.

Focusing your mind will also help alleviate the fear and panic that is natural in this situation. In spite of all the difficulties, just believe you are going to make it out of the mess you're in.

Mentally rehearse what you would do if If violence should erupt on the aircraft flight. What you would do if it happened on the ground.

Keep the floor area in front of your seat clear. If shooting should start, you'll want to duck down as low as you can for protection. If a blanket or clothing is available, grab it as you duck down and use it for extra protection.

If shooting does break out, stay down until you're convinced—completely convinced—it is over. Discipline yourself against your natural curiosity to look around. The immediate goal is to miss any flying bullets.

Keep your seat belt on tightly while airborne. Should bullets or a bomb blast cause an explosive decompression of the plane, grasp your arms under your thighs and hold on tight. Hopefully that seat belt will keep you from being sucked out of the plane if you're near the hole in the fuselage.

When things settle down, you'll realize you're isolated from the rest of the world. Worry or concern about your family may become an obsession for you. There's nothing you can do about it, so there's no use wasting too much energy with such worry.

Torture of hostages in a hijacking situation is rare. You're not military POWs with valuable information that could help in a combat situation. There is a real possibility you could end up captive in a state that allows or sponsors terrorism like Lebanon or Libya, which is what happened to TWA 847.

The young American sailor, Robert Stethem, was tragically murdered, but the other 152 captives survived. Some were released early and 40 American men were held 17 days in captivity, not too much worse for the wear and tear. If you find yourself in a similar situation, don't get downhearted. Others have survived and your chances should be excellent, too.

If you end up in solitary confinement, use your mind. It's amazing what POWs have done in such situations. They've designed and built dream houses one nail at a time. They've planned future projects, written poetry, recalled Bible verses—hundreds of things. After extended periods of time in solitary, they were even able to recall things from the past and their youth that they had long since forgotten. The human spirit and body are hard to beat—if you can keep a positive mental attitude.

If you are in a captive situation off the airplane, some type of a daily exercise program is smart. Simple calisthenics, like sit-ups, push-ups, and running in place, will help make you feel better, use up time, and keep you fit. Naturally, you should tailor your program to match your conditions.

Start a calendar to keep track of days. Set it up for two months, that way you won't be disappointed if you're not immediately released.

When you are isolated from others, communication and fellowship with your comrades becomes an insatiable craving. In Vietnam, the simple TAP code was used among U.S. POWs to communicate with each other.

Medal of Honor winner Admiral James Stockdale outlines the code effectively in his book, *In Love and War*[*]. The code is easy to memorize and could become invaluable. There's no need to memorize it now—but be familiar with how it's structured. You'll have plenty of time to get familiar with using it later, if you ever need it.

On those cold December nights, Robbie gave me lessons in how to tap messages through the wall. What had started in August as the Smitty Harris tap code (named after the prisoner who had come across it by accident during a coffee-break conversation at survival school years before) had taken on a standardized form of American Hanoi prisoner usage. Robbie instructed me to call him up with the shave and a haircut beat: tick tick ta tick tick. He would let me know he was ready to receive my first word by answering tick tick. Then I would spell it out using the five-by-five matrix I had memorized: tick tick tick tick tick (pause) tick tick (w) tick (pause) tick tick tick (h) tick (pause) tick tick tick tick tick (e) tick tick tick (pause) tick tick tick (n)—WHEN—and Robbie, understanding it, would acknowledge with a tick tick. Then my second word: tick (pause) tick tick tick tick (d) tick tick tick (pause) tick tick tick tick (o)—DO—YOU—THINK—WE—WILL—GO—HOME? After he tick-ticked acknowledgement of the HOME, I ended my message with another shave and a haircut, and he would tick-tick and then give me a shave and a haircut and start his answer—

* Published by Harper and Row Publishers, New York. (Used with permission.)

which in those days most felt compelled to answer in phrases like THIS SPRING.

Many refinements to this code had been worked out already at the Zoo (special prison compound). Robbie gave me a number of abbreviations that had become standard. For instance, on the very first day of his instructions, I learned not to laboriously spell out WHEN DO YOU THINK WE WILL GO HOME?, but to use WN DO U TK WE GO HOME? Robbie taught me to buy a word with an early tick tick as soon as I had heard enough of it to guess what it was. If I guessed wrong—i.e., if my word didn't make sense in the context of what followed—I would just have to give a series of ticks (the error signal) until I backed the sender up to where I could get it retransmitted. With constant practice and refinement, my tap communication became more accurate and almost as fast as talking.

The Smitty Harris Tap Code

	1	2	3	4	5
1	A	B	C	D	E
2	F	G	H	I	J
3	L	M	N	O	P
4	Q	R	S	T	U
5	V	W	X	Y	Z

First digit: row
Second digit: line
Examples

   S: 4-3
   T: 4-4
   O: 3-4
   C: 1-3
   K: 1-3

Besides tapping, the code was communicated with blinking eyes, whisks of a broom, clearing of the throat, and other creative ways thought out by minds desperate to communicate with their fellow man.

CHAPTER 11

What to Do in a Rescue Attempt

IF THERE IS AN ATTEMPT TO RESCUE YOU ABOARD the airliner, it will be sudden and without warning.

If it's going to be a surprise to the hijackers, it's going to be a surprise to you, too. So be prepared mentally and physically.

While being held hostage, look around and try to figure what you'd use for protection. Get this thought-out early. Be sure you know where the exits are.

When the confusion of a rescue starts, get down and get as much protection as possible. Keep quiet and still. If you're moving around, you're going to get shot.

Any sudden moves will be considered by the rescuers as a danger. Stay away from any doors that may be used for entrance.

Don't start getting involved. Your chances of being a hero are slim. Don't try to disarm your captors.

The assault team has split-second decisions to make in deciding whether or not to shoot and who to shoot. If you are found with a weapon in your hand, you could easily be killed before you have the opportunity to explain what's going on.

Let the hijackers be exposed to their fire.

Don't start yelling directions to your rescuers. They know you're there, and they have a specific plan. Don't start asking questions. Follow any directions the rescue team may shout while the action is going on. After the rescue effort is completed, you will be issued orders. Follow them.

Hijackers have often tried to escape capture by hiding among hostages. Sometimes hostages have even let them do so because of the emotional bond that sometimes forms between hijackers and hostages in the extended period of intense emotional contact. Because of this possibility, the assault force will treat everyone as a suspect until all people have been positively identified. So if you are treated roughly at this point, realize it is for your own protection.

If you are ordered to evacuate the plane through window and door emergency exits, leave all your carry-on items behind to expedite your departure and that of others. You can get them later. The plane will be barricaded off and no one is going to walk off with them. Besides, lives are more important at this stage than physical possessions. However, be sure to wear your shoes.

Get away from the aircraft as quickly as possible. There's always the possibility of fire or explosion. There will be others to help you at that point.

CHAPTER 12

If You Are Released Ahead Of Other Hostages

A TERRORIST HIJACKING NEEDS JUST ENOUGH captives to gain worldwide attention and media coverage.

Several hundred passengers on an airplane could be a control dilemma for the hijackers, especially if there are just a few terrorists in charge of many hostages. With that in mind, women, children, and some old frail men could be released early.

Such an early hostage release would leave the strong men in the hands of the terrorists and, as a result, give the terrorists the ability to extend the ordeal to achieve their demands. They would be relieved of the bothersome details of the sick and infirm.

Those released become important to the survival of the remaining hostages. Authorities will want to debrief them to learn about the actual situation aboard the aircraft.

How many terrorists are there? Where are they stationing themselves on the plane, what doors are barricaded? How many weapons? What kinds of weapons are they using? What explosives do they have? Where are the hostages seated? Has everyone been bunched up in a group? Are any sick left aboard? Did you see the pilots and what was their condition? The authorities may have many more questions.

Be as cooperative as you can. Your alert observations when you are on the plane could save lives. They will be grabbing for any grain of information they can gain from you. Be careful while the crisis is still underway. You would be wise to say absolutely nothing, except perhaps that you are happy to be freed and hope the others will be shortly.

The hijackers might request that newspapers, radio, and TV reporters be brought aboard the plane. Any information, derogatory comments, or even complimentary comments you might make about the hijackers could play into their hands or cause such alarm that the lives of the other hostages would be threatened.

The result of any loose talk could lead to a miscalculation, anger, or a false bravado by the terrorists which could prolong the event and/or endanger the lives of the remaining captives.

Just explain to the media that you'll be happy to talk with them when it is all over for everyone—but that until then, you don't want to do anything that might endanger the lives of anyone. Responsible media representatives will understand.

CHAPTER 13

After It's Over

THE HOSTAGE SITUATION COULD END IN A VARIETY of ways: assault, negotiations, or surrender. In a country of state-sponsored terrorism, the hijackers could simply release their hostages when they feel their goals have been reached.

Prior to the release, there can be a series of emotional ups and downs for the hostages. The hijackers could think the situation is about resolved, and you could become euphoric.

Then there could be a snag, and your emotions would hit rock bottom as you realize you're going to be kept captive and subject to life-threatening danger for an additional undetermined time. The emotional roller coaster could greatly affect you. This was a problem with Vietnam POWs and their release. Be cautious with your emotions until actually released and clear of any influence of the terrorists.

During the ordeal, captives become dependent upon the hijackers for survival. This period of extreme stress often creates a strange bond between the captive and captor.

The phenomenon is known as the Stockholm Syndrome. The basic symptom is sympathy by the captive toward the captor. It can inhibit hostages from being helpful to the authorities with information to properly identify and prosecute the terrorists. The syndrome generally fades with the passage of time, but it is important for ex-hostages to think with their minds rather than their hearts and emotions following release.

The length of the ordeal will have a direct bearing on the emotions of the hostages once they are released. A variety of emotions can surface. Some of the psychological and physical reactions could include:

- Guilt for lack of action to prevent hijacking
- Shame for not escaping
- Disgrace for putting others at greater risk than themselves
- Feelings of being less valuable as a person

- Critical thoughts about themselves and others for not having done enough during the crisis

- Remorse over the loss of another hostage when they survived

- Concern that others who weren't hostages couldn't understand their emotions

- Insomnia

- Nightmares

- Lack of motivation

- Erratic temperament

- Sexual adjustment

- Loss of appetite

- Feelings of alienation and detachment

- Mild phobias

All these problems are considered normal and most fade with time. The best cure is to return to your normal lifestyle and put the incident behind you.

Some hostages may experience psychological or physical reactions that require professional help. Some may have difficulty adjusting to their former lifestyle and require assistance and counseling by a mental health specialist to work out the problems brought on either as a direct or indirect result of their hostage experiences. It must be remembered that the hostage's family has also been through a very stressful period and may require the same type of assistance.

To recover rapidly from their experiences, former hostages need the support of their families, friends, and coworkers.

On many occasions, other people are uncomfortable around the former hostage because they are worried that they will not know how to act. They are afraid that bringing up the hostage incident will be upsetting. Former hostages may interpret the uneasiness of those around them as an indication of a lack of acceptance.

Generally, former hostages will eventually feel a strong need to tell their story or to share their thoughts. Others can help in this process by listening carefully to what they have to say and encouraging them when they show a desire to talk.

CHAPTER 14

Travel Agent Efforts and Concerns

THE NUMBER OF TRAVEL AGENTS HAS MUSH-
roomed since deregulation of the airlines.

In fact, the number has just about doubled,
and there are now nearly 30,000 travel
agencies in the United States.

They have one prime purpose—to serve
the travel needs of their clients, both large
and small.

In essence, their business is to see that you
have a good time, are comfortable, and have
no travel difficulties with tickets, rooms,
and tours.

About 85 percent of all international travel is booked though travel agents. Most travel agents make their living from a commission paid them by the airlines and other travel-related businesses as a percentage of the ticket price or hotel charge, etc. With airlines, that can be as much as 10 percent.

The agents need repeat business to remain in business, so they all want to do a job that you will appreciate. They know there's plenty of competition for your business if they don't.

They are also concerned about your safety and are providing special services. Those that care most are keeping country files to help you know more about the areas you are traveling to. Some are providing customers with cards giving key, helpful phrases in the language of their destination as well as with cards containing important embassy and police phone numbers for emergencies. They also provide extra copies of your itinerary for you to leave with a relative or friend so your whereabouts can always be tracked while you are away.

Some agencies even subscribe to services that provide key information about countries and give seminars on traveler safety to major companies.

Terrorist activity has caused tremendous changes in travel plans, which has taxed travel agents to the near breaking point. For example, after the TWA 847 hijacking, the U.S. Travel Data Center did a survey and reported that of the 6.5 million Americans who had made reservations abroad, 1.4 million of those had changed their reservations. That's nearly 22 percent.

Of those who changed their reservations, 850,000 people canceled their plans outright, 220,000 rebooked U.S. destinations instead, and 150,000 changed to a different foreign destination. That was a massive amount of work for agents, all without any additional compensation—but just part of their job.

The terrorism scare has cost travel agents and airlines a lot of business. However, many people are still flying and traveling.

Besides your safety as a client, what has travel agents shaking in their boots is their legal liabilities should some harm come to you from a trip the agent organized and arranged.

Most anyone would logically recognize the travel agents have no control or knowledge of any terrorist acts about to take place. Still there are questions. Much of the worry stems from several lawsuits—one from a couple booked by a travel agent on the ill-fated *Achille Lauro* cruise and another by a passenger on the TWA 847 flight hijacked from Athens.

Agents worry about leisure clients who could be influenced by an agent's advice and recommendations. "We live in a litigious society," remarked one agent.

It's recognized most business travelers travel at the request of their company, which accepts the responsibility of sending that employee. What are travel agents doing about leisure clients?

A survey by *Travel Weekly Magazine*, a travel trade publication, says 29 percent of agents are protecting themselves legally against possible lawsuits brought by clients involved in terrorist incidents.

That protection involves insurance, disclaimer forms, and verbal comments to travelers by agents. Some agents try not to urge a trip. They simply say, "I have no trouble going there myself, but I can't speak for you."

The general feeling is a need to be positive. Too many disclaimers or requets for liability forms by travel agents are not considered healthy business.

Some agents say a disclaimer could raise a red flag—perhaps discouraging a client from going at all. Some advise against such an approach because it starts the whole thing out on a negative aspect. There's got to be a better way than putting a skull and crossbones on the ticket jacket.

Most travel agents believe they should not advise clients that it is safe to travel. Final decisions must be made by the traveler.

Others claim travel agencies are relying on assurances from the airlines that any trip that looks dangerous will be canceled. Nearly everyone in the industry believes all the television and newspaper publicity gives agents protection. After all, they point out, any traveler today is aware of travel risks and of high-risk areas.

The American Society of Travel Agents, a large trade association, simply advises agents to "be of service, but don't recommend."

Perhaps Joel Ables, publisher of *Travel Trade* magazine, sums up travel agent views best:

> We should be equating the unfortunate disasters (and terrorist acts) which have occurred with facts such as the 7.7 million plus U.S. scheduled domestic and international flights which took off and landed safely during 2000.
>
> While relating the dangers of international travel to the dangers of staying at home, both the traveling public and the consumer press should be reminded of the FBI's U.S. crime statistics which point out that one's chances of being raped, robbed, burglarized or being caught in a stickup at one's local supermarket or neighborhood bar are many times higher than those of being caught in an air hijacking or any other act of terrorism.
>
> Obviously American tourists "are at risk" when traveling overseas or now in the United States.
>
> But isn't everything we do in life a risk, starting from the time we get out of bed each morning?

> The questions which we should each ask ourselves are: is it worth the daily risk of going up and down the stairs in our own homes, or of riding in the family car, or of traveling overseas for vacation enjoyment; or should we lie safely in our beds and thereby eliminate most of life's dangers, as well as its pleasures?

Some good news for agents came from a New Jersey jury that absolved a travel agency and others of any liability for booking a couple to Grenada just prior to the U.S. invasion of the tiny island in 1983.

The couple filed suit because they were caught in the middle of the fighting and were evacuated from the island by the U.S. Army. They claimed the incident caused one of them to suffer a serious illness.

The deciding factor in the trial was the couple had requested to go to Grenada on their own and were not persuaded to do so by the travel agent. The couple's attorney claimed that had the agent recommended or urged them to go to Grenada, the agent would have been held liable and the couple would have won the lawsuit.

Travel agents will play a critical role in determining whether the travel industry can counter the media hype. Agents can also play a part in keeping the public calm under the barrage of TV and print coverage.

CHAPTER 15

Living in a Foreign Land

MANY PRECAUTIONS APPLY UNIVERSALLY TO ALL who travel, and there are some concerns specific to those who spend more than a brief time in a high-risk foreign country.

A person's security ultimately depends upon his or her willingness to make some changes in lifestyle. It takes commitment.

Terrorists have changed their targets. They no longer just strike political and military officials. They are after random victims and especially businessmen abroad. Part of the change is due to the better protection of the military and diplomatic targets. Terrorists are just like anyone else. They stick to where they are most successful.

It's important that you prove to them through your actions that you are an unpredictable and unattractive target. If you are predictable in all your movements, it permits precise planning and minimizes risks for the terrorists who attack you.

When you live or work in a foreign country, times of arrival and departure at the office should be varied. Routines of travel should be constantly changed. Every day take a different route to the office and a different route home. If you are touring, take different ways to your destinations.

Stay away from any route, including jogging on the same path. The objective is to cause the would-be attackers to give up on you and select some other more predictable targets.

If you're going to live in a high-risk area for an extended period, particularly if it is for business or government work, the State Department has some excellent suggestions for many precautions you can take to protect yourself from a terrorist attack. Those precautions have appeared in numerous documents relating to terrorism, but it is useful to review those that pertain to hostage-taking situations and, to a certain extent, to assassinations as well.

While you may be familiar with many of these recommendations, a review of the safety precautions later in this chapter should be useful. You may, upon a change of circumstances, decide that a precaution that was not appropriate before is appropriate now.

Constant reevaluation is essential since the terrorist scene is dynamic. Unless you are willing to take an active part in your own defense, security professionals can do little to ensure your safety.

There are no guarantees in today's volatile world. But there is much people can do to make themselves less susceptible to terrorist violence and to improve their chances of survival should their efforts fail. Most precautions are designed to discourage terrorist attacks and not to make them impossible.

Surviving a Hostage-Taking

The Department of State has shared with their employees some guidelines in case of capture and imprisonment. We must recognize that if a terrorist organization is determined enough, if it has the resources and the backing, and if it is willing to take the increased risk for a possible greater gain, it can carry out an attack regardless of the precautions you have taken. These factors make it necessary to discuss how you can increase your chances of surviving a hostage-taking situation.

There are basically three forms of hostage taking: kidnapping, hostage barricade, and hijacking. The difference between kidnapping and a hostage barricade is that in a kidnapping the location of the hostage-takers and the hostages is unknown, whereas in a hostage barricade their location is known and the hostage takers are using the hostages as a form of protection or exploitation.

On some occasions a kidnapping may be turned into a hostage barricade situation if, during the negotiation, the location of the hostages becomes known. Many aspects of hostage survival are the same for all forms of hostage-taking. Any differences will be made clear.

A hostage-taking can be divided into phases in many different ways, but for the purposes of discussing survival the most useful division is as follows: capture, transport or consolidation, and holding and termination.

Capture

The moment at which a person is captured is very traumatic no matter where that capture occurs—in the home, in an automobile, in an airplane, or in the office.

You may be suddenly transferred from a relaxed and complacent frame of mind to a state of absolute terror. This is a tremendous transition to which people may react instinctively in a variety of ways. Some freeze, while others automatically put up resistance. It is important that you gain control over your emotions as rapidly as possible so that you can react in ways that are calculated and rational and that are designed to increase your chances of survival.

The terrorists are generally well-armed and thoroughly trained for the type of attack that they are undertaking. They are also under a great deal of stress, which may cause them to react violently or with overwhelming force to the slightest sign of resistance. Even if you put up no resistance, the terrorists may use some violence and, as a result, cause some minor injuries, because their adrenaline is flowing and they are trying to establish firm physical and psychological control as rapidly as possible.

Any sudden or misinterpreted movement could be very dangerous. They are undoubtedly hostile toward you and/or the country that you represent. They have not had the opportunity to get to know you as an individual, so they will feel no compunction about killing you if it serves their purposes to do so. At this point of the attack, except in the case of a hijacking, the terrorists always have the option of simply assassinating their victims and escaping if things seem to be going wrong.

This option may suit their purposes nearly as well as the hostage-taking would have. Naturally, it is in your interest to ensure that they do not see this option as the most attractive in your case.

During an armed assault on a group such as would normally be found at an office or at a reception, stay low and be as inconspicuous as possible. Do nothing to draw attention to yourself.

If someone else in your group puts up resistance or attempts to escape, shooting may result. To avoid being hit by stray rounds, seek whatever cover is available. At a minimum you should lie flat on the floor.

Generally, the best reaction for most people is to concede that the terrorists have the upper hand and that resistance would be futile and dangerous. You must work toward regaining a rational frame of mind and avoid any impulsive behavior.

Putting up no resistance when resistance would obviously be counterproductive will help you to avoid injury and avoid the destruction of personal property that may be important to you, such as eyeglasses or other items that you need for your comfort and long-term survival.

While your captors may want to keep you alive and will often provide medical attention when you need it, such attention may not be readily available. Even relatively minor injuries can become quite serious when untreated in possibly unsanitary conditions.

In general, you should follow immediately any instructions that are given to you and consider the possible consequences of anything that you do. Above all, do not make yourself appear threatening.

At some point, most people consider avoidance of capture or escape. There are undoubtedly times when escape is possible, but it requires an unlikely combination of events. Some of the factors that should be present before you attempt escape are:

1. You are in excellent physical condition and prepared for the rigors normally associated with escaping from an armed group.

2. You are mentally prepared for the attack and ready to react before your captors have consolidated their position.

3. You have a specific plan that is suited to the situation.

4. You have been trained to use special driving techniques. This training is available for a fee at various driving schools in the United States.

If you take your potential captors by surprise with a well-thought-out and well-executed escape plan, you may have a good chance of evading them. In most situations, however, attempts at evasion are futile. And in any situation, there will always be a point at which resistance is counterproductive no matter how well-trained or prepared you are.

The terrorists will move quickly to gain control over you. Once they have done so, you are unlikely to be given a good opportunity to escape. This is not to say that you should not be alert to such opportunities, but you must evaluate them carefully. It is senseless to risk your life needlessly if history shows that most victims are released unharmed.

Most people go through a range of psychological reactions immediately after being taken hostage, but the most common responses are fear, denial, and withdrawal. Some have experienced such overwhelming fear that they began to question their ability to cope with the situation. Fortunately, most of us are able to cope with much more than we had ever expected.

One way that the mind deals with this fear is to deny that this experience is actually happening to them. In essence the mind is unable to accept the information that the senses are providing to it. Others have withdrawn into themselves, effectively shutting off any external input. One form of withdrawal is to fall asleep. There are reports of some hostages who have slept for as much as 24 hours while being held captive.

While our specific reactions may vary, the reassuring thing is that we are able to cope mentally with a great deal of stress, and we show great resilience in recovering from extremely trying events.

Transport or consolidation

It is at this phase that there is a wide divergence among kidnappings, hostage barricade situations, and hijackings.

In the case of kidnappings the victims are moved to a secret location. Certain aspects of this move have particular significance for survival. In the case of a hostage barricade or hijacking, the terrorists, shortly after the capture of their victims, work toward consolidating their position so that they are less vulnerable to attack from the outside. Naturally, terrorist activities vary and, therefore, the victim response will vary.

In the case of a kidnapping, terrorist groups move the victim as quickly as possible from the scene where he or she was seized. During the capture, the victim will probably be handled roughly, and this rough treatment is likely to continue, partly due to the nervousness and haste of the terrorists, but also due to an attempt to intimidate the victim and gain immediate submission. Some captives are made unconscious either by a blow to the head or through the use of drugs.

Generally, it is not advisable to resist attempts by the terrorists to administer drugs, because the drug used is generally harmless and designed to make the victim more manageable and less aware of his or her surroundings. In fact, the drugs may be beneficial to those having difficulty regaining control over their emotions or to those who are in pain. The drug may produce a calming effect or induce sleep. In either case, the drug may relieve a stressful transition with a minimum of mental or physical trauma.

Whether drugged or not, you will probably be blindfolded, bound, and gagged; forced into a position that may be awkward or painful; and placed in a very confining space, such as in a shipping trunk, the floor of the back of a car under the feet of your captors, the trunk of a car, or even in a sack.

Feelings of claustrophobia may result, but it is important to remember that this confinement is temporary. You will probably be released shortly from this extreme confinement to an area that, while possibly still very confining, is more tolerable.

You will undoubtedly be very frightened and at least somewhat disoriented during transport, but you must work toward regaining your composure. It is important that you use this time to gain whatever information you can through a very active use of whichever senses you have available to you.

Listen, for example, for tire sounds that may indicate what type of road surface you are traveling on; for outside noises such as factory sounds, traffic noises, or aircraft; for voices that may enable you to subsequently identify your kidnappers; and for distinctive motor or other vehicle sounds. Attempt to sense from the motion of the vehicle any turns in the road and other topographical features.

Note, also, any unusual odors that might indicate certain types of manufacturing activities, a fishing industry, the nearness of the sea or other body of water, the presence of particular types of vegetation, or other types of activities or locations that have distinctive smells associated with them.

Some hostages have even managed to leave a trail by dropping something or making a mark in any location that they are placed so that their presence in that location can later be proved. Be extremely cautious in doing this, however, since being caught in the act could lead to severe reprisals.

In a hostage barricade situation, the activities of the hostage-takers will be different. Their prime interest in the early stages of the incident will be to fortify their position.

As in a kidnapping, you may be treated roughly, drugged, blindfolded, bound, and/or gagged. It is also possible that you will be confined in a small space or have to endure long periods of time in uncomfortable positions. You may, on the other hand, be forced to help the hostage-takers in moving furniture or otherwise making the site less vulnerable to attack.

Some hostages have even been used as human barriers by being placed near windows or doors to further discourage an assault.

As with a kidnapping, it is important that you begin to collect information immediately.

Often, some hostages are released earlier than others. If you are selected for early release, you must take advantage of this opportunity to provide to the authorities any information you have memorized concerning the terrorists, the site, and the other hostages. Such information is vital for the negotiations and for any possible rescue attempt.

You must also identify a location that provides you with some cover or concealment in the event of an armed assault. There have been occasions in hostage barricade situations where the authorities assaulted immediately without a period of negotiations.

In general, then, during either transport or consolidation your most important functions are to regain control over your emotions and activities and to gather intelligence. The quicker you do this, the greater your chance for survival.

After your captors have moved you to a holding location for the ensuing negotiations, or after they have consolidated, you will enter what will probably be the most lengthy phase: holding.

Holding and termination

At the beginning of this phase, your captors probably consider you as just an object that will be useful to them in their bargaining with the authorities. They do not consider you a human being with needs and concerns, and they will do things that are designed to maintain that view. Their activities are intended to make you more compliant and to make it easier for them to kill you if they find it necessary.

They may, for example, make accusations that you are a member of an intelligence organization or that you were engaged in activities that were detrimental to the people of the country. They may attack your political point of view while advocating their own, and they may attempt to force you to do things that are degrading or insulting.

Often they will try to isolate you from events by placing you in a darkened cell or otherwise denying you all sensory input, removing your sense of time, and refusing to provide you with any information about what is happening in the outside world.

Sometimes they will cause you to doubt that anything is being done to gain your release. Some organizations may even resort to attempts to convince you of the validity of their goals, they may threaten you, or they may try to intimidate you with their knowledge of your previous activities or lifestyle.

Occasionally hostages are kept in an unsanitary environment, dehumanized by being called by a number or some designation other than their name, provided inadequate or no hygiene or bathroom facilities, deprived of sleep, and given poor or insufficient food.

The specific type of treatment can vary dramatically. What is typical is that your needs may go unrecognized, partly because of their determination not to recognize you as a human being and partly because of their animosity toward what you represent. While they may make life unpleasant for you, they are unlikely to do anything intentionally that is life-threatening.

Dead, you are worth nothing to them. As time goes on, you can certainly make use of this fact. You may never be treated in the way that you would like to, but you may be able to bring about some modification of their behavior.

During this phase you should continue trying to cope with your new environment and the behavior of your captors and with the thought that you may not be released soon, particularly in kidnapping situations. Depending upon early release may prevent you from doing those things that are essential to your long-term survival, and it can cause you to lose all hope as time wears on.

Perform four main functions during this time:

- Build human relations
- Manage yourself
- Manage time
- Manage your environment

Building a relationship with your captors does not mean that you have to agree with their philosophy or their methods. It means simply working toward establishing a relationship that permits you to maintain your self-respect while demonstrating that you are a valuable, nonthreatening human being.

Present yourself as a reasonable and intelligent person. Thoughtless displays of emotions, whether it be aggression or panic, can cause your captors to see you as a threat or as a person who is worthy only of contempt.

If you are given orders, comply with them unless they are life-threatening or exceedingly degrading. If you feel it is better not to follow a particular order, explain why you are resisting it. Sometimes they may understand and respect your reasons.

Avoid arguments with or criticism of your captors. Even if you win the argument or feel some personal victory from a well-placed insult, you will have made some dangerous enemies on whom your life may depend.

If your captors show some willingness to talk, encourage this and listen attentively.

Show interest in their point of view even if you do not agree with it.

Take any opportunity to talk about yourself, your hopes, your aspirations, your failures, and your worries. The more your captors know about you, the more difficult it will be for them to kill you.

Managing yourself involves setting goals for yourself, organizing your activities, and establishing some control over what it is that you are going to do.

It is important for both your sense of mental well-being and physical health that you try to maintain your personal cleanliness and hygiene to the extent that your environment permits it. Some hostages have even been given the materials necessary to make repairs to their deteriorating clothing. This helps you not only maintain your sense of self-respect, but it also provides an activity to help pass the time.

Establish a physical exercise program and a relaxation program. Physical exercise helps to keep your body in good condition, and it can be very relaxing by giving you a sense of accomplishment and a healthy tiredness that improves sleep. Many relaxation techniques are effective in reducing stress levels or the effects of stress on the body.

It is well known today that stress, particularly if it is maintained over a period of time, can have adverse effects on your health. Keep your mind active by reading whatever literature is provided to you and using any writing materials that might be available. If these materials have not been provided, ask for them. You may not be given the kinds of reading materials you would have chosen for yourself, and you may not be permitted to send or even retain any of your writings, but what is important is that the mind is kept active.

In the absence of reading or writing materials, you can still engage in memory exercises or in imaginary problem solving. Some hostages have drafted poetry or prose in their minds and were able to write out these materials word for word after the incident was over. Others have imagined undertaking a project for which they had to work out all the details. Many creative uses of the mind cannot be denied to you no matter how severe the environment.

Those who are religious have received great comfort from their religious beliefs, and in many cases those beliefs became stronger because of their experiences. Some have even felt that without those beliefs they might not have been able to survive their experiences.

Try to maintain your sense of humor and an optimistic outlook on life. A sense of humor is probably the most effective way of warding off apathy and depression.

Managing your time requires that you be aware of the passage of time and that you establish a regular schedule for all your events. In many cases your captors will attempt to remove your sense of time by removing your watch, keeping you in an area that deprives your sense of input, and refusing to answer any questions about the time or date.

In spite of their efforts, however, you can determine approximate time by detecting changes of temperature that are associated with day and night; observing the patterns in which meals are served; detecting outside noises such as bird sounds, traffic noises, and factory noises that traditionally occur at certain times of the day; and watching the behavior of the guards who may appear more sleepy and less active at night than during the day.

Establish a calendar for yourself so that you can keep track of the passage of days. Some hostages have been so effective at this that they were only a few hours off in their calculations after months of captivity. Plan for a long captivity by establishing a calendar of at least 30 to 60 days and then extending it if necessary, rather than having unrealistic expectations of a short stay dashed.

Be prepared for disappointments. Unexpected complications may arise, creating lengthy delays in what you think is an impending release.

Keep track of holidays, birthdays, and other special days so you can celebrate.

Establish a schedule for yourself for each day and stick to your schedule to the extent possible. Your schedule might include times for personal hygiene; housekeeping; exercise; relaxation; reading, writing, or mental exercises; special work tasks such as mending clothing or other repairs; and sleep. A day that is active and full passes much more quickly and is more personally satisfying.

You can't always control when you will be fed, but you can control when you eat. Stockpile food and water for the possibility that a meal will be missed or the food that is provided is inedible.

Managing your environment involves making the space in which you are confined your personal space or your temporary home. There are a variety of ways in which this can be done. You can rearrange those items of furniture that are movable into a pattern that is more pleasing to you. You can designate specific areas for specific events such as setting aside a corner for exercises and another corner for reading and writing. You can keep your space neat and organized. And you can request additional pieces of furniture that you feel you need. You may not get all that you ask for, but this should not discourage you from trying. Any additions to your possessions can be very important to your sense of accomplishment and pride.

Interrogation   While a captive, you can expect that some form of interrogation will take place.

It will range from mild questioning stemming primarily from curiosity to very sophisticated interrogation by those who are seeking information that would be damaging to the interests of the United States or cause severe embarrassment.

Fortunately, however, most of the interrogation conducted by terrorists is not sophisticated and is, therefore, more easily resisted. It is important for your personal sense of integrity and self-respect that, to the extent possible, you remain loyal to your country and to any companions.

Some of the methods commonly used by interrogators are:

- ○ The appearance of being friendly
- ○ Shock and surprise
- ○ Fear and despair
- ○ Playing the role of the noninterrogator
- ○ Threat and rescue
- ○ Accusations of spying
- ○ Electronics
- ○ Disgrace

Good interrogators will start out by appearing to be friendly because the victim is looking for a friend. The interrogator may treat you so well in comparison with the others that you are grateful and want to do for them whatever they request. Some hostages even feel guilty when they cannot remember some seemingly insignificant piece of information that the interrogator is asking for.

An interrogator may use threats in a friendly way by saying that if they don't come up with useful information, they will be forced to turn you over to someone who uses harsher measures.

Interrogators use shock and surprise by revealing, for example, that they already know a great deal more than you would have thought. You may feel that if they already know so much, then the damage is already done, and what little you could tell them will not make much of a difference.

Interrogators will also know what you fear most. They may know, for example, that one of your primary concerns is for the welfare of your family, and they will play upon this fear to induce you to talk.

Interrogators have often posed in another role, such as a neutral party who is just checking on the welfare of hostages. U.S. government employees, and some businessmen, are often accused of being spies. You may be inclined, therefore, to provide information about your actual duties to prove that you are not engaged in spying.

A wide variety of interrogation techniques could be used. And this is why it is important that you remain alert to any attempts to obtain information from you.

Whenever you are asked a question, pause before answering to determine where the question might be leading. There is a second reason for establishing the habit of pausing. People generally pause on sensitive subjects but not on nonsensitive subjects. Pausing on all questions denies the good interrogator the opportunity to find out about what subjects you are sensitive.

It may be tempting to provide information in order to improve your condition, but you must consider the consequences for your country and for your own mental well-being.

Some of the more common methods of coercion are solitary confinement, inadequate or unsavory food, and depressing living conditions. Solitary confinement can take many forms. It is not necessary to be confined to a dark cell. Being blindfolded, having stereo headphones placed over your ears, and being kept in a small tent can have the same effects.

The early stages of solitary confinement are generally the most difficult because you have a great deal of time to kill mentally. Eventually, however, people become very creative. Some people concentrate on trivia. Others solve math problems. Some review past events, books they have read, or movies they have seen. In the absence of distractions people can recall past events in amazing detail. Some hostages have taken imaginary trips or built complex structures board by board. This sort of creative use of the mind is available to all of us.

Most hostages have reported that it was difficult or impossible for them to maintain adequate sanitary conditions. Cleanliness is very important to us in our culture, but, in truth, you have no real need for a daily bath, deodorants, and soaps. Use these items if you have them, but do not be concerned if you do not. You may be offensive to those around you, but that is their problem, not yours. It is unlikely to have any effect on your health.

The major concern for most people is the possibility of the use of torture or the systematic application of pain. The use of torture to gain information, however, is not generally very effective. What an interrogator needs is useful, reliable, detailed information, but a hostage who has been mistreated is often less capable of providing it.

The sophisticated interrogator, therefore, will generally not rely on torture, but rather use the techniques referred to above.

If torture is used, however, how do you deal with it? Normally, if a source provides no information, the interrogator will give up and concentrate on someone else.

Most people, though, cannot remain silent in the face of torture. So, you need some alternatives. You can make it evident, first of all, that you are not in a position to have any information that would be useful to them. This is your preferred tactic, but it may not always be possible if others have already provided information about you or classified documents have already been compromised.

Your next alternative is to make it clear that you have information, but you are not able to give it in a useful, reliable manner.. You can claim that you do not remember information, and considering the stress you are under, that would be understandable; that you are unable to think clearly because of an injury you suffered when taken, perhaps a blow to the head; or that you cannot understand their questions because of language difficulties or accents.

If you are forced through torture to reveal information, that certainly is understandable. But force them to do it again for each piece of information they desire. When you provide information use idiomatic language and unusual words to make yourself difficult to understand, be vague and indefinite, be misleading and inaccurate, and digress into irrelevancies.

If they see that they are not going to obtain useful information from you, they may leave you alone. Many people find that they can tolerate much more than they thought they could.

Above all, it is important that you behave in ways that will preserve your sense of integrity and self-respect. This will greatly improve your recovery from the hostage-taking once it is over.

Precautions Overseas

Here are some suggestions for individual and family precautions; residential, office, and transportation security; and recognizing surveillance.

Individual precautions

○ Memorize emergency telephone numbers such as police, embassy or consulate, security personnel, and medical assistance, and carry change or tokens for pay telephones. Learn how to use the local telephone system.

○ Do not publish home addresses, telephone numbers, or personal information concerning family members.

○ Be alert to surveillance. (See the discussion on surveillance later in this chapter.)

○ Learn the local language at least well enough to be able to communicate an idea or a need for assistance, no matter how ungrammatically correct.

○ Prepare an itinerary and keep your family and office informed of your whereabouts. Advise your office of your arrival and departure and of any changes in your plans.

○ Obtain maps of the area and become familiar with the city, concentrating in particular on your neighborhood, the neighborhood in which your office is located, and routes to and from your office. Locate on the map all areas where emergency assistance can be obtained.

○ High-threat targets should discuss with security professionals the need for bodyguard protection and the size of the protective detail required. To be effective, bodyguards must be adequately equipped and trained.

Precautions for the family

○ Make sure that family members receive appropriate briefings and that they are aware of the security threat in your area.

○ Have family members memorize telephone numbers for emergency situations, and make sure that all family members, particularly children, know how to use the telephone. Post emergency telephone numbers on or near the home telephone.

- Warn family members against revealing information about travel or other family plans.

- Avoid discussing personal information over the telephone. Children in particular should be told not to provide information over the phone since their natural inclination is to provide whatever information is asked of them.

- Warn children about being approached or questioned by strangers.

- Where appropriate, ensure that children are properly escorted to and from school or other events by a parent or guardian.

- Advise your entire family to vary daily routines such as shopping trips, social gatherings, and family outings.

- Have all family members report any unusual or suspicious event immediately to security personnel.

- Have all family members advise other members of the family of their plans for the day. In high-threat situations it is advisable for family members to report their arrival at and departure from each location and to inform other family members of any change in plans.

- In high-threat areas, advise family members to travel in groups as much as possible by coordinating shopping and other trips with other families.

○ Advise family members to be cautious in allowing strangers to enter the home. Verify that any service people who request entry have a legitimate reason for being there by contacting the person who requested their services.

○ Develop a specific plan as to what to do in the event of a terrorist incident or other security emergency.

Residential security

○ Where possible, have your residence surveyed by a professional security officer. Otherwise, get guidance from the embassy or your company's security officer on what residential security standards should apply for your area, and make an information survey of your own residence.

○ Get a recommendation from the embassy or your company's security office as to what security hardware is recommended or suitable for use on your residence.

○ If you are selecting your own residence, look for housing in an area where other diplomatic residences are located, where there is easy access to arterial roads, and where there are no other facilities which might attract crowds.

Ideally, the residence should have a secure perimeter, a lockable garage, and adequate exterior lighting.

Having residences close together permits sharing of guard services and permits arrangements for responding to alarms.

○ Be alert to persons disguised as public utility crews, road workers, vendors, etc., who might station themselves in places to observe potential targets and collect information needed to plan an attack.

○ Note the license number of suspicious vehicles and the description of their occupants. Report such incidents to the security officer or local police.

○ Be aware of vendors, inspectors, repair people, or investigators who seek entry. Anyone who is unknown or unexpected should not be allowed to enter, no matter how official they may sound or appear, until you have satisfactorily established their identity.

It is preferable to screen visitors using a peephole or intercom before opening the door.

○ If your residence is equipped with a radio on an emergency net, make sure all members of your family, including children, know how to use it.

- Select a safe room which will provide the family with temporary protection until help can arrive. The room should have a sturdy door with a lock and an emergency exit. Check with your security officer for recommendations as to what other emergency equipment should be stored there.

- Develop a family emergency escape plan. Each member of the family should have an emergency escape route through a clearly designated and specially prepared exit.

- Become familiar with the capabilities and limitations of the local police. If additional protection is needed, consider obtaining a residential guard. Remember that guards must be trained and properly equipped to carry out the responsibilities that you assign to them.

- Consider requesting police protection for large social events.

- Where possible, obtain a background investigation on domestic employees before employment, or, at a minimum, request references and biographical data. Contact former employers and listed references. Brief domestic employees on security practices.

- If an intruder is detected, move the family to the safe room and call for assistance. Remain in the safe room until help arrives, unless you are forced to abandon it through your emergency exit and escape route.

○ Consider installing an intrusion alarm.

Office security

○ Do not provide the location or travel plans of employees over the telephone. The travel plans of all employees should receive restricted distribution.

○ Home addresses and telephone numbers must not be given to unknown people or to telephone callers.

○ All employees should be alert to people disguised as public utility crews, road workers, vendors, etc., who might station themselves near the office to observe activities and gather information. They should also watch for parked or abandoned vehicles near the entrance or walls.

○ Provide license numbers and descriptions of suspicious vehicles and descriptions of their occupants to the embassy or company's security officer.

○ Take seriously your responsibility for escorting visitors, and be careful about whom you bring into your office building.

○ Become familiar with all emergency plans and participate in any drills

that are held. Know the appropriate
action to take when alarms are activated. Know all escape routes and
the location of safe havens and emergency equipment.

Transportation security

○ Vary your times and routes of travel.

○ Where possible, use different doors
  and gates when arriving at or leaving
  the home or office. It is also advisable to have an automatic gate or
  garage door opener or an employee
  who can open the gate for you as
  you approach. Many attacks take
  place when a person is entering or
  departing the residence because this
  part of the route cannot be easily
  altered.

○ By using alternate entries and depar-
  tures to and from the residence and
  office and avoiding unnecessary
  delays, you can significantly reduce
  your vulnerability.

○ Whenever possible, leave your car in
  a secured parking area. Where no
  secure parking is available, vary your
  parking space so that your route
  between your car and the office or
  residence is varied.

○ Be especially alert in underground
  garages. Check the area before exiting the vehicle.

○ If you see something unusual, drive away.

○ Always check the vehicle, inside and out, before entering it. If you notice anything unusual, do not enter the vehicle.

○ Before leaving a building, check the streets for suspicious individuals or vehicles.

○ Be familiar with your route and all alternate routes.

○ Whenever possible, use well-traveled routes, avoiding remote areas or areas where the traffic could easily be blocked.

○ If you must travel in remote areas, travel in groups and keep others informed of your travel plans.

○ Know the location of police, hospital, military, government, public facilities, and other secured areas along your normal routes that could be used as safe havens in the event of an attack.

○ Avoid using vehicles that identify you as an American.

○ Maintain your vehicle in good operating condition and keep the gas tank at least half full.

○ Consider installing a vehicle alarm that would protect your vehicle from tampering or intrusion.

- Where permitted, a two-way radio or telephone installed in the car is an excellent security precaution.

- Keep your vehicle doors locked and the windows closed.

- Drive defensively, keeping alert to your surroundings and leaving adequate space around your vehicle for emergency maneuvering. Check regularly for surveillance (see the special section on this subject). Look ahead for existing or potential roadblocks.

- Change or reverse direction to avoid being boxed in. In high-threat situations you should constantly imagine what you would do if an attack were to take place. This mental preparation can save precious seconds if an attack actually occurs and would probably result in a more effective response.

- Always wear your seat belt in case you need to take evasive action in your car.

- If danger is suspected, report it using the vehicle radio or telephone, if you have one, and drive immediately to the nearest appropriate safe haven.

  If there is shooting, lower yourself in the vehicle so that you present less of a target and depart the area as rapidly as possible. Most rounds will pass through the body of an unarmored vehicle

But they may be deflected and their velocity and penetrating power will be reduced. If your vehicle is armored, know the capabilities and limitations of the armor.

○ If you are driven by a chauffeur, ensure that the chauffeur has been adequately briefed and, if possible, trained by security specialists in evading or countering terrorist attacks.

Surveillance

Just about all training programs designed to protect individuals from becoming victims of terrorism recommend that people be alert to surveillance. This is excellent advice, but unfortunately, in most instances it is insufficient, because people have had no training in detecting surveillance, and terrorist organizations are often relatively sophisticated in their surveillance methods.

Surveillance detection conducted by trained experts is not as easy as most Hollywood films would lead us to believe. Fortunately, however, the type of surveillance conducted by terrorist organizations is not normally as elaborate as that done by intelligence organizations, nor does it involve as many people or as much equipment. Nevertheless, for people to have a reasonable chance at detecting most forms of surveillance, they would have to be somewhat familiar with the techniques used.

The purpose of surveillance is to determine (1) the suitability of the potential target based upon the physical and procedural security precautions that the individual has taken and (2) the most suitable time, location, and method of attack. This surveillance may last for days or weeks depending upon the length of time it takes the surveillants to obtain the information that they require. Naturally, the surveillance of a person who has set routines and who takes few precautions will take less time. The people undertaking the surveillance will often not take part in the attack nor will the attack take place while surveillance is still in progress.

Before undertaking surveillance, most experts gather all information that is available about the subject from other sources. Public records or information made available to the terrorist organization from a sympathetic individual within the embassy, local police, or other government office may reveal useful facts about an individual such as the names of family members, an address, descriptions of vehicles and license numbers, photographs, etc. The surveillants will also make a reconnaissance of the neighborhood in which the target lives and works. This permits them to select positions of observation, the types of vehicles to use, the clothing to be worn, and the type of ruse to use that will give them an ordinary or normal appearance and a plausible reason to be in the area.

There are basically three forms of surveillance: foot, vehicle, and stationary. A brief description of the most common techniques used for each of these forms of surveillance and the methods for detecting them follows.

Foot surveillance may be undertaken by one or more individuals. One-person foot surveillance is rather complicated and fairly easy to detect by an adult person. The surveillant must remain relatively close to the target, particularly in congested areas, to avoid losing him or her.

In less congested areas the surveillant can maintain a greater distance, but the lack of other pedestrians makes the surveillant that much more noticeable. The one complicating factor is the use of a disguise to make the surveillant look different. One indicator of the possible use of a disguise is a shopping bag or some other container for a change of clothes, particularly if the shopping bag is from a store not found in the area or the container somehow seems out of place. Where a disguise is suspected, pay particular attention to shoes and slacks or skirts. These items are less easily and, therefore, less commonly changed.

In elevators, watch for people who seem to wait for you to push a button and then select one flight above or below yours.

Two-person foot surveillance is more effective in that the second surveillant provides greater flexibility. Normally, one surveillant remains close to the target while the other stays at a greater distance. The second surveillant may follow the first on the same side of the street or travel on the opposite side. Periodically the two surveillants change position so that if the target spots one of them, that one will soon be out of sight, leading the target to think that he or she was mistaken. Obviously, spotting this form of surveillance is more complicated, but individuals who are alert to the people in their vicinity will eventually detect the same surveillant over a period of time.

Foot surveillance with three or more people uses the most sophisticated techniques and is the most difficult to spot. Generally, one surveillant remains behind the target close enough to respond to any sudden moves. A second surveillant remains behind the first on the same side of the street with the first surveillant in sight. A third surveillant travels on the opposite side of the street parallel with or just behind the target.

In areas where the target has few paths to choose, one surveillant may walk in front of the target, where he or she is least likely to cause suspicion. The positions for the surveillants are frequently changed, most commonly at intersections. The surveillant directly behind the target may move to the opposite side of the street, while another surveillant moves in close behind the target. With the additional surveillants, any surveillant who feels that he or she has been observed may drop out of the formation. The use of this sophisticated technique requires that people be alert not only to those people behind them but also to those across the street and perhaps in front of them. If the same person is seen more than once over a certain distance, surveillance may be suspected even if that person is not continuously seen.

Common methods for detecting surveillants apply to all three forms of foot surveillance. Some of the most effective are:

1. Stopping abruptly and looking to the rear

2. Suddenly reversing your course

3. Stopping abruptly after turning a corner

4. Watching reflections in shop windows or other reflective surfaces

5. Entering a building and leaving immediately by another exit

6. Walking slowly and then rapidly at intervals

7. Dropping a piece of paper to see if anyone retrieves it

8. Boarding or exiting a bus or subway just before it starts

9. Making sudden turns or walking around the block

While taking these actions, watch for people who are taken by surprise, react inappropriately, suddenly change direction, or give a signal to someone else. Surveillants will not normally look directly at the target, but they may do so if they are surprised or unaware that you are observing them.

Foot surveillance is often used in conjunction with vehicle surveillance, since it is likely that the target will use a combination of foot and vehicle transportation. Vehicles used for surveillance are generally inconspicuous in appearance and of a subdued color. Frequently, the inside dome light is made inoperative so that it will not illuminate the interior of the car when the door is opened. Vehicles will have two or more people in them so that if the target parks his or her vehicle and walks away, the surveillance can be resumed on foot while the driver remains with the vehicle. While moving, the driver gives full attention to driving while the observer operates the radio, watches the target, and makes notes on the target's activities.

Sometimes it will be necessary for surveillants to break traffic regulations to avoid losing you. If you see a vehicle run a red light, make an illegal U-turn, travel over the speed limit, or make dangerous or sudden lane changes in an apparent effort to keep up with you, you should, of course, be suspicious of that vehicle. The distance between a surveillance vehicle and the target will vary depending on the speed at which the vehicles are traveling and the amount of traffic. In most cases, however, surveillants will try to keep one or two vehicles between them and the target.

As with foot surveillance, vehicle surveillance may be undertaken using only one vehicle or using two or more vehicles. One-vehicle surveillance suffers from the same drawbacks as one-person foot surveillance. The target has to be kept in view at all times and followed by the same vehicle. Surveillants can try to overcome this disadvantage somewhat by changing seating arrangements within the vehicle; putting on and taking off hats, coats and sunglasses; changing license plates; and turning off onto side streets and then turning back to resume the tail. This makes it necessary for a person suspecting surveillance to remember aspects of a following vehicle that cannot easily be changed such as the make, model and color of the car and any body damage such as rust, dents, etc.

The use of two or more vehicles permits surveillants to switch positions or to drop out of the surveillance when necessary. One vehicle follows the target vehicle and directs other vehicles by radio. The other vehicles may follow behind the lead vehicle, precede the target vehicle, or travel on parallel roads. At intersections, the vehicle following directly behind the target vehicle will generally travel straight ahead while alerting all other vehicles of the direction in which the target has turned.

Another vehicle in the formation will then take a position behind the target and become the lead vehicle, taking over the responsibility for giving instructions to the other surveillants. The former lead vehicle then makes a U-turn or travels around the block to take up a new position, ready to resume the lead vehicle position again when necessary.

People who have well-established routines permit surveillants to use methods that are much more difficult to detect. If, for example, you leave the office at the same time each day and travel by the most direct route to your home or if you live in a remote area with a few or no alternate routes to your home, surveillants have no need to follow you all the way to your residence.

Two alternative methods of surveillance in such situations are leading surveillance and progressive surveillance. In leading surveillance the surveillant travels in front of the target while the observer watches for turns. When the target turns, this is noted. The next day the surveillant makes a turn where the target did the previous day. Over a period of time the surveillants will discover the entire route to the residence while still driving in a position that creates much less suspicion.

There are two forms of progressive surveillance. In the first form, surveillants are placed at intersections along the probable route of the target. When the target makes a turn, this is noted and the position of the surveillants is adjusted to check the next intersection. Eventually, this method leads the surveillants to the residence. In the second form of progressive surveillance, a vehicle will follow the target for a short distance and then turn off. On successive days the surveillant picks up the target where he or she left off the previous day. Leading and progressive surveillance are extremely difficult to detect, but you should not give anyone the opportunity to use these methods.

The most effective methods for detecting most forms of vehicle surveillance are:

1. Making a U-turn where it is safe to do so

2. Making a turn to the right or left (in general, left turns create greater complication for surveillants because of oncoming traffic that might delay a turn)

3. Going through a traffic light just as it is turning red

4. Stopping just beyond a curve or hill

5. Circling a block

In each case, watch for the reactions of any vehicles that you may suspect. Any vehicles that make unusual maneuvers should be carefully noted. Do not forget to check for motorcycles or motorbikes, since in many parts of the world they seem to be favored by surveillants because they move easily through heavy traffic.

Stationary surveillance is probably used most commonly by terrorist organizations. As mentioned earlier, most attacks take place near the residence, because that part of the route is least easily varied. Most people are more vulnerable in the morning when departing for work, because morning departure times are more predictable than are evening arrivals.

Surveillants seek a position that permits them to observe the residence or office clearly without being observed or suspected. If the surveillants decide that it is best not to be seen, they may obtain an apartment in the area, which provides an adequate view, but such apartments may not be available and the renting of an apartment could provide clues for a subsequent investigation. The use of an apartment for surveillance, while possibly the most difficult to detect, is generally not the easiest or safest method.

Many surveillance teams use vans with windows in the side or back that permit observation from the interior of the van. Often the van will have the name of a store or utility company to provide some pretext for its being in the area. The driver may park the van and walk away, leaving the surveillance team inside. Some teams use automobiles for stationary surveillance, parking the vehicle far enough from the residence or office to be less noticeable, using other vehicles for cover, facing the vehicle away from the target, and using the rearview mirrors to watch.

Where it is not possible to watch the residence unobserved, surveillants must come up with a plausible reason for being in the area. The types of ruses used are limited only by the surveillant's imagination. Some of the more commonly used covers are automotive repairs due to engine trouble or a flat tire, door-to-door sales, utility repair crews, lovers in a park, walking a dog, construction work, or sitting at a café. Women and children are often used to give a greater appearance of innocence.

Some things to check for are parked vehicles with people in them, cars with more mirrors or mirrors that are larger than normal, people seen in the area more frequently than seems normal, people who are dressed inappropriately, and workers who seem to accomplish nothing.

If you become suspicious of a van, note any information printed on the side of the van, including telephone numbers. Check the telephone book to see if such a business exists, or call the number. Note the license numbers of any suspicious vehicles and provide them to your security office so they can check. Make a habit of checking the neighborhood through a window before you go out each day. Consider photographing any unusual individuals or activities, discreetly if possible.

Detecting surveillance requires a fairly constant state of alertness and, therefore, must become an unconscious habit that you form.

We do not want to encourage paranoia, but a good sense of what is normal and what is unusual in your surroundings could be more important than any other type of security precaution you could take. Above all, do not hesitate to report any unusual event. Many people who have been kidnapped realized afterward that their suspicions had been well-founded. If those suspicions had been reported, their ordeal might have been avoided.

Conclusion

While there are no guarantees that these precautions, even if religiously adhered to, will protect you from terrorist violence, they will undoubtedly reduce your vulnerability and, therefore, your chances of being selected as the next victim.

You owe it to yourself and to those who care for you to do those things that are required of representatives of the U.S. government or companies in a foreign land.

Certain risks are associated with many professions, so you should be prepared to take the steps that those risks force upon you.

Many of the victims of past terrorist incidents have been found to have ignored even the most basic security recommendations. Perhaps if they had taken the risk more seriously or had greater faith in preventive measures, they might have avoided their life-threatening encounter.